BOB HARRIS' *Guide to*

CONCRETE OVERLAYS & TOPPINGS

This book is dedicated to all of the wonderful students and customers we have had the pleasure of associating with through the years. We cherish the relationships we have built with many of you that will last a lifetime. Without all of you as a source of inspiration, the field of decorative concrete would be otherwise boring, dull and lacking of excitement! Keep up the great work and use each and every day as a learning experience.

Thank you to all of the manufacturers for the many products and accessory items that have contributed to the success of our workshops as well as other professional workshops in the industry. Without your fine products, educators and installers would not have such a diverse medium to not only express themselves with, but earn a living at doing something we all love, decorative concrete!

Thank you to Lee Ann for all of the contributions as well as the hours and hours of editing required for this book. (Someday we will take our show on the road, honey, in the Kingsley Coach!)

International Standard Book Number: 0-9747737-2-7

Printed in the United States of America

First Printing: December 2005

Credits

Publisher	Jim Peterson, ConcreteNetwork.com, Inc.
Associate Producer	Lee Ann Stape, Decorative Concrete Institute, Inc.
Executive Editor	Anne Balogh, ConcreteNetwork.com, Inc.
Design	Christina Wilkinson, Sabre Design & Publishing
Publicist/Marketing	Khara Dizmon, ConcreteNetwork.com, Inc.
Photography	**Front Cover:** (top) Skookum Floor Concepts, Ltd.; (middle) Decorative Concrete Institute; (bottom) Bay Area Concretes (Photo courtesy of Bomanite Corporation) **Back Cover:** (center) River Alloy Designs (Photo courtesy of Colormaker Floors); (top) Concrete FX; (middle) Modello Designs; (bottom) Rudd Company

Special thanks to Jim Peterson, Khara Dizmon, and Anne Balogh for your extreme professionalism and intense motivation. Without such an inspired team, these guides would not exist. It is scary to imagine what all of us working as a team have accomplished. (Stay tuned for more.)

Finally, I have to thank my entire family for being there when we need you the most.
I know you are watching from up there, Dad. This Bud's for you!

Bob Harris' Guide to Concrete Overlays & Toppings is the third in a series of guides for construction professionals on popular decorative concrete topics to be published by the Decorative Concrete Institute, Inc. and ConcreteNetwork.com, Inc. This and *Bob Harris' Guide to Stained Concrete Interior Floors* and *Bob Harris' Guide to Stamped Concrete* form the Bob Harris Decorative Concrete Collection.

ABOUT THE AUTHOR

Bob Harris is known worldwide in the decorative concrete industry.

He is president of The Decorative Concrete Institute, an organization in Douglasville, Georgia, that offers hands-on training and workshops in the latest decorative products and techniques.

He also was affiliated with a large manufacturer of decorative concrete products for nearly a decade, the last part of which he served as the director of product training. In this role, he conducted hands-on training seminars in architectural concrete in locations around the world in addition to being an integral part of technical support and research and development.

Central to Harris' ability to teach is his extensive experience with the products he trains contractors to use. Prior to his director of product training role, he worked as senior superintendent for six years exclusively in Disney theme parks, doing decorative "themed concrete" work with integral color, dry-shake color hardeners, and chemical stains. On the Disney properties, he did everything from casting tree branches to alligator paw prints in concrete.

He personally placed and/or supervised the placement of over three million square feet of decorative concrete, including work for some of the major Disney theme parks in Orlando, Florida.

Exceeding expectations is what lights Harris up. It could be when an owner he produces a floor for says, "This work exceeds my wildest expectations," or when he teaches a contractor to build floors that will elicit the same elated response from the owners that contractor works for.

Harris also shares his expertise with others through his involvement with industry associations. He holds three certifications from the American Concrete Institute and serves on ACI Committee 303, Architectural Concrete; ACI Committee 610C, Field Technician Certification; and ACI Committee 640C, Craftsmen Certification. He also is affiliated with the American Society of Concrete Contractors and is a member of ASCC's Decorative Concrete Council, as well as the author of several articles for various technical publications.

BOB HARRIS

INTRODUCTION

If you're fortunate enough to work in the decorative concrete industry like I do, you've experienced firsthand the explosive growth in the demand for our product.

More and more people tell me they want decorative concrete, but many also admit they can't afford to replace their existing concrete with a gorgeous new pool deck, patio, driveway, or floor. Now they have an alternative, thanks to recent technological advances in overlays and toppings designed to rejuvenate concrete floors, exterior flatwork, and even wall surfaces. When existing concrete needs a face-lift, these systems can be a cost-effective solution to the dreaded remove-and-replace option.

Overlays and toppings can be applied at thicknesses of 1/4 inch or less yet result in an extremely durable, aesthetically pleasing surface. An endless array of decorative options in color, pattern, and texture permit complete design flexibility for makeovers that are totally unique.

The growing demand for these resurfacing systems, by both the residential and commercial markets, offers a huge opportunity for decorative concrete contractors who are willing to learn the techniques and have the desire to be creative. My purpose with *Bob Harris' Guide to Concrete Overlays & Toppings* is to share the knowledge I have gained from my experiences to provide you with a solid foundation for producing quality work and to help you avoid common mistakes and problems.

Here are some of the topics covered in-depth in this guide:

- A to Z instructions for installing underlayments, microtoppings, spray-down systems, stampable and self-leveling overlays for industrial, commercial, and residential slabs (and even walls), both indoors and out.
- How to choose one application versus another.
- Budget analysis to determine how much to charge for each type of application.
- How to estimate materials.
- Surface preparation and crack repair.
- How to deal with and test for moisture-vapor transmission.
- Coloring options, including color layering and the use of stain, dyes, tints, and broadcast colors.
- Decorative finish options, including the use of stamping mats and stencils, scoring techniques, working with gelled acid, and polishing.
- How to set up a mixing station.
- Labor requirements for each technique.
- Curing times required prior to walking on, sealing, or stamping the work.
- Troubleshooting pointers for each type of application.

You'll also find a list of the essential tools, equipment, and supplies needed to mix and apply each type of system as well as an impressive pictorial gallery of completed projects.

Although this guide can certainly help you sharpen your skills and shorten your learning curve with overlays and toppings, it is not intended to be a substitute for on-the-job experience—the only way you can truly master these methods. My goal is to give you a comprehensive understanding of the fundamentals while introducing you to the advanced techniques being used by seasoned pros. Enjoy!

TABLE OF CONTENTS

Almost indistinguishable from natural stone, a vertical overlay is an economical alternative for enhancing a fireplace. Special stamps produce deep faux grout joints while stain accents impart realistic color detail.

CHAPTER 1

CHOOSING ONE SYSTEM OVER ANOTHER

When it comes to choosing an overlay system, one size does not fit all. Each system described in this guide has unique qualities—with distinct performance criteria to match. With the many options available, you'll probably find at least one system that meets your client's requirements for durability, appearance, maintenance, and cost. But how do you choose the ultimate system for the job?

In this chapter, I discuss some of the key factors to consider. Be sure to go over these issues with your clients as well, so they can participate in the decision-making process. By informing them of the advantages and the limitations of each system, you will help to ensure the selection of a product that lives up to their expectations.

Condition of the existing surface

The condition of the existing concrete slab will be one of the biggest considerations when choosing the appropriate system. Does the concrete exhibit excessive cracking or deep chips and gouges? Is the surface out of level? If so, the slab would not be a good candidate for a microtopping or skim coat designed to be applied at a thickness of 1/8 inch or less. These imperfections, if not perfectly patched or filled, could telegraph through a thin topping. After repair, a floor like this would be a better candidate for a higher-build material, such as a self-leveling or stampable overlay. (To review procedures for crack repair, read Chapter 10.)

Moisture content of the concrete

Too much moisture in the concrete substrate can be a problem for many overlay systems, especially products that restrict the ability of the concrete to "breathe," or release moisture as needed. As the moisture vapor travels up through the capillaries of the concrete and reaches the surface, it can cause the topping or overlay to delaminate.

There are several causes of excess moisture in concrete. If you suspect you have a problem, read Chapter 8 to learn about a couple of simple tests you can perform to check the moisture-vapor emission rate. If the rate is high, check to see what level is acceptable for the system you want to install. Some overlays will permit the moisture vapor to escape more readily than others.

Durability

Resurfacing systems permit you to experiment with a wide range of artistic effects, but your creative artwork is all for naught if the overlay is vulnerable to chipping, abrasion, or chemicals. Although many of the systems described in this guide stand up well to wear, chemical attack, and dirt penetration when properly installed and maintained, some systems are better suited than others for harsher exposure conditions. Often a higher-build material will perform better in aggressive environments.

Be sure to ask the facility owner specific questions about the type of conditions the resurfaced concrete will be exposed to. Is the slab outdoors and subject to weather extremes and seasonal freezing and thawing? Is it a floor surface in a high-traffic area or in the line of fire for chemical, grease, and oil spills? Will heavy equipment or furniture be dragged across the floor? Make sure the topping or overlay system you choose is tough enough to withstand all the conditions it will confront. Often you can find this information in the technical data sheets for the product (see Chapter 12).

Desired look

Of course, the main reason to use a resurfacing system is to give the existing concrete a face-lift. While all of these systems will certainly make the surface look better, each type has its own distinct appearance. The processes for achieving color, pattern, and texture also differ among systems (see table comparing the attributes of the various overlay systems on page 13). I discuss these decorative possibilities in greater detail in the chapters devoted to each technique.

A stamped overlay in a random stone pattern contrasts beautifully with hardwood flooring and carpet.

CONCRETE SOLUTIONS, INC.

Cost

The main factors affecting the cost of materials are the applied thickness and type of coloring method used (such as integral coloring, broadcast pigments, stains, or dyes). The total cost of installation will also be higher if you are using stencils, stamping mats, or decorative sawcutting to pattern the surface. For a detailed breakdown of the average installed costs for each system, turn to Chapter 4.

Maintenance requirements

Clients often have great expectations about their newly resurfaced concrete floors early on, but then neglect to maintain them properly and are disappointed when the surface ends up looking dull or lackluster. For any installed topping, the amount of maintenance required is a function of the amount and type of traffic the floor will receive. In most cases, routine sweeping to remove any abrasive particles and occasional wet mopping with a mild detergent or soap will prevent damage to the finish. Textured or patterned surfaces may require more aggressive scrubbing with a bristle brush or mechanical floor sweeper to remove dirt or grease that becomes trapped in the depressions in the surface. Occasional pressure washing may be needed to clean exterior surfaces.

As you can see, choosing the most appropriate resurfacing system often involves weighing performance attributes along with aesthetic value. Chances are, there's a system available that will strike the perfect balance for the job at hand.

Self-leveling overlays are the perfect canvas for sawcut patterns, with stains providing accent color.

With stamped overlays, stunning stone-like texture and color can be achieved on a limited budget.

A microtopping adds flair to commercial environments, and permits customization with logos and other graphics.

In a residential entryway, a microtopping offers the upscale look of marble while giving homeowners an easy-to-clean floor surface.

A COMPARISON OF SYSTEM TYPES

System	Application Thickness	Decorative Options	Performance Attributes
Underlayments	Featheredge to 1 1/2 inch	Not applicable	Restores structurally sound yet worn concrete in preparation for floor coverings. Provides a smooth, level base.
Self-leveling overlays	1/4 to 1 inch (in one lift)	Can be left seamless (except at joints) or used as a canvas for sawcut or engraved designs. For color, mix in integral pigments or apply stains, dyes, or tints.	Restores structurally sound yet worn concrete surfaces and can be used to level uneven floors.
Microtoppings or skim coats	Featheredge to 1/8 inch	Can be troweled to a texture or troweled silky smooth. You can incorporate color into the mix, broadcast dry pigments onto the surface while troweling for a marbleized appearance, or accent with stains and dyes. To achieve variation, try layering different colors with each successive lift.	Because these systems are polymer-based and applied super-thin, they have the ability to flex, unlike other systems. May also be applied to wall surfaces.
Spray-down systems	1/8 inch	Usually applied as a splatter coat or a knock-down (troweled) finish, and are ideal for use with paper or plastic stencils. Available precolored or can be integrally colored during mixing. Apply stains, dyes, or tints for accent color.	The textured finish is durable and skid-resistant, making it well suited for surfaces such as pool decks, sidewalks, and driveways.
Stampable overlays	1/4 to 3/4 inch	Can be stamped and textured to mimic brick, slate, natural stone, and other materials. Color options include broadcast pigments, colored liquid or powdered releases, stains, dyes, and tinted sealers.	Are applied thick enough to accept texture and cover minor imperfections in the concrete substrate.
Vertical stamped overlays	Up to 3 inches (in multiple lifts)	Go on thick enough to permit the creation of deep rock textures and other designs using stamps or hand sculpting techniques. Accent coloring with stains or dyes makes it possible to reproduce the multi-toned, weathered look of natural stone or aged brick.	Specially formulated to be applied to primed vertical surfaces without sagging. Also are "carveable," so you can achieve stone or masonry wall patterns with deep reveals and grout lines.

A microtopping with a seamless texture mimicking worn stone forms a curved pathway to guide shoppers, giving them an experience similar to hiking outdoors.

Decorative sawcuts and a combination of stains, dyes, and tints give this trowel-applied microtopping a distinctive look fit for a gourmet food store.

CHAPTER 2

FINISH AND TEXTURE OPTIONS

Different textures and finishes are possible with each of the overlay systems described in this guide, ranging from ultra-smooth to heavily textured. With some imagination, the right tools, and a bit of experimentation, there really is no limit to what you can achieve. Described below are just a few of the possibilities. For more details about achieving each finish, turn to the chapter describing the overlay application where it's commonly used.

Knock-down finish

A knock-down texture is probably the most popular finish for spray-applied toppings (see Chapter 16). To produce this finish, installers typically use a trowel right after spray application to "knock down" the high spots left behind from the sprayer, leaving a slightly textured surface. Some professional installers create a variation of a knock-down finish by dipping a brush into the mixed topping and splattering the material onto the surface

The other option is to simply leave the texture as is after spray application of these toppings. The hopper gun produces a distinctive texture similar to the irregularly shaped domes of a spray-applied acoustical ceiling. But the more subtle texture of a knock-down finish is more comfortable to walk on. This makes it a good choice for surfaces intended for bare feet, such as pool decks.

Textured skim coats

With skim coats or microtoppings (covered in Chapter 17) you can also achieve a variety of textured finishes. One method is to dribble on the second coat by reaching into a bucket of mixed material with a trowel or spatula and letting the material randomly splatter onto the floor. Then, before the material begins to set, trowel the raised areas back down to level, in a fashion similar to the aforementioned knock-down technique. You can make this your final texture, or you can take it one step further by trowel applying one more

coat of material, using a finer mixture possibly in a different color. This fine skim coat fills the voids left behind from the previous coat, leaving interesting textural variations similar to those found in natural travertine.

Because these systems are finished with a variety of tools, such as trowels, floats, or rubber squeegees, you can vary the texture by simply modifying your troweling technique. For example, holding your hand trowel flat as opposed to cocked when applying the skim coat leaves a light texture similar to smooth slate.

The use of a rubber float instead of a trowel will give you a rougher, sand-like finish ideal for surfaces requiring good slip resistance, while a magic trowel or rubber squeegee will produce a relatively smooth surface.

Broom finish

Many of my clients like the classic look of a broom finish, which produces a corduroy-like effect ranging from subtle to pronounced, depending on the coarseness of the broom used. Some manufacturers offer overlay systems specifically designed for creating broomed textures. Some standard microtoppings can also be finished in this manner.

Generally, the broom finish is one of the easier textures to obtain. Once the substrate has been prepared per the manufacturer's recommendations, the topping can be applied by a squeegee, trowel, or broom. The surface is usually ready for final broom finishing 5 to 15 minutes after

topping application (depending on jobsite conditions). A fine- to medium-bristled broom typically used for brooming conventional concrete slabs works best. Small hand brushes or paint brushes are handy for working in tight areas, such as 90-degree corners, or for brooming against curved surfaces, such as column walls.

Hard-troweled finish

Not all of your customers will want a textured finish for their overlays. Some may prefer a smooth, highly polished look. This can be achieved by going over the surface with successive passes with a steel trowel, producing a look similar to that of hard-troweled concrete. This technique is well suited for use with semi self-leveling microtoppings. Generally, this type of finish should be limited to indoor surfaces

ELITE CRETE SYSTEMS, INC.

Texturing skins produce a seamless texture with no grout lines. In addition to floors, they are also ideal for texturing steps, or in this case, a raised fireplace hearth.

because it offers little slip resistance, especially when wet.

Stamped textures

With stampable overlays (see Chapter 18), the stamping mats and texturing skins produce patterns and textures that can mimic materials such as stone, brick, or slate. Texturing skins produce a seamless texture with no grout lines while mats produce a deeper pattern with well-defined lines. On many projects the two are combined, such as using texturing skins for steps or accent borders and stamping mats for an adjacent walkway or patio.

Flat, smooth surface

Unlike some of the other toppings described in this guide, self-leveling overlays generally produce a smooth, seamless look with no texture (depending on the applicator's skill level). The material is first gauge raked to the proper depth and then smoothed with a steel smoothing blade. Read Chapter 15 for more details on installing these overlays.

Decorative scoring

Decorative hand scoring and sawcutting can be used with most overlays to create unique patterns and custom graphics. There are many different tools available for decorative scoring and cutting. My favorites include:

- An angle grinder with a 4-inch blade.
- A Crac-Vac—a 7-inch upright grinder with a vacuum attachment for dust-free cutting of straight lines and crack routing.
- The Mongoose—a hand-operated sawcutting machine that makes fast, accurate straight cuts as well as perfect circles.
- The Wasp—an air-driven engraving tool for highly detailed areas.
- A Dremmel with a diamond tip, for detail work and sections with very tight radiuses.
- Handheld scoring tools for carving lines.

(Note: You can find some of these tools at hardware or art supply stores or equipment rental houses. The Mongoose and Wasp are available through Engrave-A-Crete. The Crac-Vac is manufactured by International Surface Preparation. See the list of resources at the end of this guide for contact information.)

Regardless of your scoring tool of choice, it is important to make sure the overlay has had time to cure sufficiently. Chipping at the edge of the sawcut could occur if you cut too soon.

Cement-based toppings lend themselves to a wide array of patterns and textures. Combining different finishes on the same project produces attention-grabbing contrast.

Applying a microtopping in multiple layers creates interesting textural variations that can't be duplicated by any other surfacing material.

A wash of neutral stain and a high-gloss sealer enhance the subtle beauty of this lightly textured overlay.

CHAPTER 3

ESTABLISHING EXPECTATIONS WITH THE BUILDER, ARCHITECT, AND OWNER

If your marketing campaign is working (see Chapter 23), it will drive a steady stream of prospective customers to your door. You've enticed these prospects by telling them about the many benefits of concrete resurfacing systems. Or maybe one of your recent projects caught their eye and they were "wowed" by your work. Now you must convince them that you have the skills to achieve the same phenomenal results on their project.

Often you will be approached by people who have already priced other decorative paving or flooring materials but want to explore the alternatives you have to offer. Whether you are dealing with a homeowner on a small residential project or a team of people (architect, general contractor, building owner, etc.) on a large commercial installation, it is important to be informative but realistic.

Now—before any contracts are signed—is the time for you to clearly explain what is and is not possible with decorative overlays and to help customers determine the most suitable system for their application. By providing a solid foundation of information early on in the project, you can build good customer relationships and avoid numerous headaches.

Explain the variables

Establishing expectations starts with your initial sales call. Discuss the differences between what your proposed system has

COLORADO HARDSCAPES

Seeing is believing. Show your clients actual installations or representative samples of the various patterns and textures you offer.

Stenciled overlay designs can be impressive additions to your portfolio. Obtain literature from stencil suppliers to show your clients the many pattern options available.

A good camera and notebook are as essential on a decorative resurfacing project as your materials and application tools. Keeping detailed written and photographic records not only allows you to retrace exactly what you did on each job, but can also be useful in defending against legal claims. Chronicling your work starts with the first set of samples you submit to the client. If you submit multiple samples that have been fabricated with different colors and patterns, label the back of the each sample and date it so you can easily identify it later.

Once the project begins, note the materials you used and in what amounts (including any coloring agents and sealers) and the curing times required between coats and before sealer application. Also note the specific colors used and the names of your material suppliers.

Supplement your notes by taking sequential photographs of the entire project—preferably with a digital camera, so you can easily post the photos on your website or organize them in a computer database. Many consumers don't understand the overlay process. Showing them photos of each phase—from surface preparation to overlay application to final sealing—can help you explain the steps involved.

Photos can also help prevent lawsuits. Before the project begins, take pictures of the existing surface and note the surface preparation and repair procedures required. Also take pictures of the jobsite and any nearby structures, especially if other trades are on the project. If a home or building owner later refuses to pay for the work, claiming that your crew failed to prepare the surface properly or damaged the surrounding walls or landscaping, you will have photographic evidence to the contrary.

to offer compared to other products being considered, such as natural stone, high-end tile, or carpet. Make buyers aware that decorative overlays are not a premanufactured good over which you have total control. Often you will be dealing with imperfect situations (such as a substrate in need of repair or major profiling) and variable jobsite conditions, especially on outdoor projects. That means total continuity and consistency of the final product will be difficult if not impossible to achieve. As with any custom handcrafted product, slight variations in the final appearance can be expected and are often desirable. Other factors, such as weather conditions, can also affect the performance and appearance of the final product.

Keep a portfolio of your work

For clients who haven't seen your work firsthand, a portfolio showing pictures of your best efforts can be a powerful sales tool. Of course, you can't create a portfolio without taking jobsite photos to chronicle your work. I have found that taking sequential installation photos helps to procure more jobs. That's because I can show prospective clients each step in the process, so they understand what's involved and can witness the transformation as it unfolds (see Chronicling Your Work on this page).

Provide samples

Once clients see all of the possibilities, you can work with them to pinpoint a specific design and color scheme that suits their tastes. At this point, you should explain to customers the differences in the looks attainable with various coloring techniques (described in Chapter 20) and decorative methods (see Chapter 1).

Color charts or sample chips can be helpful here, and some manufacturers will provide them at no charge for use as a sales tool to assist in color and material selection. Keep in mind, though, that the color charts are on paper, not on concrete, and most of the sample chips are small in size and made in a controlled environment. In other words, they won't realistically represent what an entire project would look like, so use them only as a starting point to determine the desired products and colors to use.

Once clients have narrowed down their options, make up actual samples showing some of the combinations they've chosen. You can do this by applying the materials to cement backerboard, commonly found in the tile section of most major home improvement stores. Giving clients something they can see and touch, rather than just visualize, will convey the true look and texture of the finished surface (see Tips for Preparing Samples on page 21).

Even better, make up a variety of samples beforehand to showcase the different systems and color selections you offer. Also consider opening a showroom or professional design center to showcase your work, as discussed in Chapter 23.

Get it in writing

Aim for the top in establishing expectations and building that great relationship with your clients—but don't forget to cover your back with the contract. The final decisions you agree on with your clients must be backed up in writing, as emphasized in Chapter 5. When working for a general contractor (GC) on a large commercial project, make sure the contract includes provisions to protect your interests, not just those of the GC and the client. It's also a good idea to have your attorney review the contract before you sign.

The condition of the existing substrate and other jobsite conditions will affect the outcome of your topping installation. Explain to your customers that minor variations and inconsistencies are to be expected.

Before and after photos of your work will give clients a visual image of what's possible with decorative overlays while showcasing your craftsmanship.

ELITE CRETE SYSTEMS, INC.

Tips for Preparing Samples

Taking the time and effort to prepare representative samples of your work is one of the best ways to ensure client satisfaction. Here are some tips to get you started:

• Make sure your samples accurately represent what you can accomplish in the field. A common mistake I see contractors make is to prepare samples that are too small to adequately represent the full effect. I recommend making your samples at least 2x2 feet in size so you have the working room to incorporate any sawcut designs or stenciled or stamped patterns you plan to use. This will give clients a better sense of what the final surface will look like.

• When preparing your sample boards, apply the materials in the exact fashion you would on the actual job. And don't skip any steps, such as coloring or sealer application, or substitute a different product. This can result in an inaccurate representation of the final installation. Document the formulas used to prepare the sample, and keep accurate notes in a job file so you can replicate what you've done.

• Don't go overboard by using all sorts of elaborate coloring techniques and spending hours to make the sample perfect, because you'll be expected to achieve the same level of perfection over the entire project. It's nearly impossible to be so meticulous and flawless on a larger installation that's a couple of thousand square feet or more.

• On large commercial jobs, you may be asked to submit a series of samples or do an actual mockup on the jobsite. Obviously, this can become time consuming and costly.

On the other hand, it's better to find out during the sampling stage if you meet the owner's expectations. I recommend factoring the costs of producing these samples into your bid. Some contractors provide the first sample free, but charge for additional samples.

CHAPTER 4

BUDGET ANALYSIS

Establishing prices for decorative overlays and toppings requires careful analysis. No one job will be identical to another, so you need to account for all the variables and adjust the cost as necessary. The goal, of course, is to charge enough to make a decent profit without setting your prices so sky high that you drive away business. Following are some of the key factors to consider when pricing your work.

System type and complexity

The type of system you're installing and the level of decorative detail involved will have the biggest influence on project cost. The table on page 25 provides average price ranges for each system type. As you can see, the cost often doubles or even triples when advanced decorative techniques are added, such as stenciled designs with color layering or intricate sawcut patterns. Not only do jobs with greater artistic complexity require more skill to execute, they often necessitate a larger investment in time, labor, materials, and equipment.

Market demand

Local market conditions will often influence what you are able to charge for overlays and toppings. If you're in an established market with an influx of competing decorative overlay installers, chances are the average market prices will be lower. Conversely, if your market is new to decorative toppings and the competition is scarce, you may be able to charge more because you are offering a distinctive, handcrafted product.

Often you will be competing against other flooring materials, such as high-end tile, marble, or granite. Depending on the complexity of the overlay installation, your price may be the same or even higher than some of these materials. This is when you need to emphasize the unique colors, textures, patterns, and finishes possible with overlays and toppings—artistic effects that simply can't be achieved with other building materials. (For other marketing tips, turn to Chapter 23.)

Project size and accessibility

Accessibility will also help to determine fair pricing, especially on large commercial or high-rise jobs. Can you set up your mixing station close to the installation area and close to the drop-off point where materials are delivered? The less material handling involved, the lower your installed cost.

As an example, my crew worked on a project where the installation took place on the 37th floor of a high-rise building. The entire floor (over 16,000 square feet) was to receive a self-leveling overlay. We had to transport over 600 bags of material weighing 55 pounds each, along with the necessary equipment and tools, the entire 37 stories up! We spent two days just mobilizing tools, materials, and equipment to the mixing station. Certainly when pricing a project like this, you must account for the extra time and expense involved. Often you will need to set the price substantially higher than what you would for a smaller project where accessibility is not an issue.

Surface preparation requirements

The prices given in the table do not include expenses for surface preparation. This can often be the most time consuming and labor-intensive phase of an overlay

BOGGS CONTRACTING GROUP; PHOTO COURTESY OF BOMANITE CORPORATION.

A complex project involving three different textures, each defined by a different accent color, will be at the upper end of your pricing scale.

For clients on a budget, stick with a one- or two-color microtopping, possibly enhanced by simple decorative scoring. The results will still be impressive.

A decorative overlay with an ornate stenciled border and color layering will command a premium price due to the extra detail work involved and the cost of materials.

project, so the amount of work required will greatly impact your final cost.

Generally, surface preparation averages between $0.75 to $1 per square foot. But that all depends on the size of the job and the degree of profiling required, such as shotblasting vs. just a light sanding or acid etching.

Repairing cracks, especially if they are structural in nature, can really eat into your project budget. You may end up charging anywhere from $1 to $8 per lineal foot, depending on the severity of the repair. Guidelines for analyzing cracks and typical repair methods are provided in Chapter 10.

Designs with elaborate patterns and multiple color variations will drive up the cost of a decorative topping because they demand a larger investment in time, labor, materials, and equipment.

COMPARING THE COSTS

Application	Price per square foot installed*
Underlayments	$1 to $3 Material cost with primer: $0.50 per square foot at 1/8 inch thickness (price may be lower on large projects)
Self-leveling overlays	Basic installation: $5 to $7 With decorative scoring and use of stains or dyes: $7 to $12 Material cost with primer: $2.75 to $3 per square foot at 1/4 inch thickness
Microtoppings	Basic one-color skim coat: $5 to $7 Multiple coats and color layering with stains or dyes: $7 to $10 Advanced applications with multiple colors and complex sawcut designs or decorative stenciling: $10 to $15 Material cost with primer, if applicable: $0.90 to $1.10 per square foot at 1/8 inch thickness
Wall veils (microtoppings applied to vertical walls)	Basic installation: $10 to $15 Material cost: Same as horizontal microtopping
Spray-down systems	Basic knock-down finish with sealer: $2.75 to $4 Basic stencil design: $4 to $6 Advanced with multiple stencil designs and color variations: $6 to $9 Material cost: $0.50 to $0.90 per square foot at 1/8 inch thickness
Spray-applied, vertical	Basic spray texture on vertical surface: $3 to $5 With multiple colors and patterns: $5 to $15 Material cost: Same as horizontal spray-down
Stamped overlays	Basic imprint with seamless texture skins: $5 to $7 Using one stamp pattern: $6 to $8 Advanced with multiple patterns, textures, and combination of coloring systems: $8 to $12 Material cost: $2.25 to $3 per square foot at 3/8 inch thickness
Vertical stamped	Basic using one to two colors and one stamp pattern: $10 to $20 Advanced using multiple patterns, colors, and textures: $20 to $40 Material cost: $2.50 to $3 per square foot at 1/2 to 3/4 inch thickness

*These budget numbers include acrylic sealer and floor finish where applicable. Add an additional $0.60 to $0.75 per square foot if protecting the overlay with a high-performance coating. Surface preparation costs are not included in these figures.

CHAPTER 5

WRITING A FAIR CONTRACT

Protecting the Floor During Construction: Who's Responsible?

On very large projects, it's nearly impossible for a general contractor to shut down the entire operation while you perform your work. This often forces you to install the overlay in phases, increasing the odds that other construction trades could inflict some type of damage to your handiwork, leading to time consuming and costly rework.

When working on such projects, clarify who is responsible for protecting the finished floor while other trades proceed with their work. If the responsibility is yours, be forewarned that the labor and material involved to protect a finished floor can be costly, especially on large jobs. Some preemptive measures:

• Perform your work toward the end of the construction process to reduce the possibility of damage to your finished topping by construction equipment and materials being dragged across the floor.

• Have a principal from the GC you are working for sign off on each area you complete before you move on to another section of the building. Then if damage occurs, you have proof that it didn't result from improper installation. Repairs to this damage can be time consuming and costly to fix (see Chapter 7 for more information).

I have seen many startup and even experienced overlay installers get burned because nothing was put in writing on important issues relating to the project (such as those covered in Chapter 3). They were only discussed verbally or not at all, leading to misunderstandings and customer dissatisfaction, or in extreme cases, a ruined reputation and costly litigation. You can avoid all of these problems by getting a signed contract, one that's fair and covers all the bases.

Such a contract starts with a well-detailed proposal, which you should present to the customer prior to being awarded the project. On a residential job, the client usually signs the proposal and the proposal itself becomes the contract. On large commercial projects where you are hired as a subcontractor, you will usually sign a contract with the general contractor (GC), rather than the owner. Sometimes the verbiage in such contracts is written to protect the GC and their client. To protect yourself, I advise having an attorney review any contract you are proposing to sign.

I read one contract, for example, that held the floor resurfacing contractor liable for repairs to any problems with the floor for an unlimited time frame after completion.

In another case, the contract stated that in the event of early termination or suspension, the contractor would be responsible for returning the floor back to its original condition—which would be impossible if the concrete had been mechanically abraded in preparation for the overlay. Always read the fine print before you sign!

Here is what a well-detailed proposal addresses:

• Location of the project
• Who the contract is between (the owner, general contractor, owner's agent, etc.)
• A complete description of services to be provided, from surface preparation to final sealing

SAMPLE
of A General Construction
Contract Form

AGREEMENT: as of the _____ Date:_____

RE: PROJECT ADDRESS: _____

BETWEEN _____

(hereinafter called the "owner") whose mailing address is:

_____

(hereinafter called the "contractor") whose mailing address is:

_____

CONTRACT DOCUMENTS

Contract Documents, which constitute the entire agreement between the Owner and the Contractor and are as fully a part of the Contract as if attached, are enumerated as follows:

(Strike through any that are not applicable to this project).

1. This "Agreement and General Conditions".
2. "Procedures for Contractors".
3. Work Write-Up and Itemized Bid Dated _____ ("Specifications and Bid").
4. General Specification Manual.
5. Addenda No. _____
6. Attached sketches/drawings.
7. Owner selection list Dated _____
8. Other _____

THE WORK

The contractor shall perform the entire rehabilitation of the residential structure as described in the contract documents except as indicated as follows to be the responsibility of others:

Scope Responsible Party

_____

TIME OF COMMENCEMENT & SUBSTANTIAL COMPLETION:

The Work shall commence within 7 calendar days of authorization by written Notice to Proceed from the Owner.

The Work shall be substantially completed no later than ___ calendar days from the date of the Notice to Proceed. The Contractor shall be liable for and shall pay the owner $_____ as liquidated damages for each calendar day of delay until the work is substantially completed.

Optional

[If Work is delayed at any time by causes beyond the Contractor's control, then the Contract may be extended for such reasonable time as the Owner's Authorized Representatives may determine.]

OWNER'S REPRESENTATIVE

The Owner's Representative shall be

_____

The Owner's Representative shall:

1. Provide administration of this Contract during construction and throughout the warranty period;
2. Visit the site at intervals appropriate to the stage of construction to determine if the Work is proceeding in accordance with the Contract Documents;
3. Based on evaluation of the Contractor's invoices for payment, determine the amounts owing to the Contractor;
4. Have authority to reject Work that does not conform to the Contract Documents;
5. If the Contractor fails to correct defective Work or persistently fails to carry out the Work in accordance with the Contract Documents, by a written order, may order the Contractor to stop the Work, or any portion thereof, until the cause for such order has been eliminated.

CONTRACTOR'S RESPONSIBILITIES

The Contractor shall supervise and direct the Work, using his/her best skill and attention, and he shall be solely responsible for all construction means, methods, techniques, sequences and procedures and for coordinating all portions of the Work under the Contract.

The Contractor warrants to the Owner that all materials and equipment incorporated in the Work will be new unless otherwise specified, and that all Work will be of good quality, free from faults and defects and in conformance with the contract Documents. All Work not conforming to these requirements may be considered defective.

The Contractor shall give all notices and comply with all laws, ordinances, rules, regulations, and lawful orders of any public authority bearing on the performance of the Work, and shall promptly notify the Owner's Representatives if the Drawings and Specifications are at variance therewith.

The Contractor shall be responsible for all safety precautions in connection with this Work. He shall take all legally required and reasonable precautions for the safety of all employees on the Work and other persons who may be affected thereby.

Contractor's liability insurance shall be purchased and maintained by the Contractor to protect him from claims under workers' or workmen's compensation acts and other employee benefit acts, claims for damage because of bodily injury, including death, and from claims for damages, other than to the Work itself, to property which may arise out of or result from the Contractor's operations under this Contract, whether such operations be by himself or by any Subcontractor or anyone directly or indirectly employed by any of them. This insurance shall be written for not less than any limits of liability specified in the Contract Documents, or required by law, whichever is the greater, and shall include contractual liability insurance applicable to the Contractor's obligations under this Section. Certificates of such insurance shall be filed with the Owner prior to the commencement of the Work.

The Contractor shall not employ any Subcontractor to whom the Owner's Representatives or the Owner may have a reasonable objection. The Contractor shall not be required to contract with anyone to whom he has a reasonable objection.

CONTRACTOR "HOLD HARMLESS" WARRANTY

To the fullest extent permitted by law, the Contractor shall indemnify and hold harmless the Owner and the Owner's Representatives and their agents and employees from and against all claims, damages, losses and expense, including but not limited to attorneys' fees arising out of or resulting from the performance of the Work, provided that any such claim, damage, loss or expense (1) is attributable to bodily injury, sickness, disease or death, or to injury to or destruction of tangible property (other than the Work itself) including the loss of use resulting therefrom, and (2) is caused in whole or in part by any negligent act or omission of the Contractor, any Subcontractor, anyone directly or indirectly employed by any of them or anyone for whose acts of any of them may be liable, regardless of whether or not it is caused in part by a party indemnified hereunder.

Such obligation shall not be construed to negate, abridge, or otherwise reduce any other right or obligation of indemnity which would otherwise exist as to any party or person described in this Section. In any and all claims against the Owner or the Owner's Representatives or any of their agents or employees by any employee of the Contractor, any Subcontractor, anyone directly or indirectly employed by any of them or anyone for whose acts any of them may be liable, the indemnification obligation under this Section shall not be limited in any way by any limitation on the amount or type of damages, compensation or benefits payable by or for the Contractor or any Subcontractor under workers' or workmen's compensation acts, disability benefit acts or other employee benefit acts.

CORRECTION OF WORK

The Contractor shall promptly correct any Work rejected by the Owner's Representatives as defective or as failing to conform to the Contract Documents, whether observed before or after Substantial Completion and whether or not fabricated, installed or completed, and shall correct any Work found to be defective or nonconforming within a period of one year from the Date of Substantial Completion of the Contract or within such longer period

of time as may be prescribed by law or by the terms of any applicable special warranty required by the Contract Documents. The provisions of this Article apply to work done by Subcontractors as well as to Work done by direct employees of the Contractor.

CHANGES IN THE WORK

The Owner, without invalidating the Contract, may order Changes in the Work consisting of additions, deletions, or modifications, the Contract Sum and the Contract Time being adjusted accordingly. All such changes in the Work shall be authorized by written Change Order signed by the Owner's Representatives and the Contractor.

CONTRACT SUM/PROGRESS PAYMENTS

The Owner shall pay the Contractor for performance of the Work, subject to additions and deductions by approved Change Orders, the Contract Sum of $_____ _____. The Contract sum is determined as follows:

Base Bid _____

Addenda _____

Contract Sum _____

Based upon invoices submitted to the Owner's Representatives, the Owner shall make payments on account of the Contract Sum to the Contractor as follows:

Draw 1 _____ % $ _____

Draw 2 _____ % $ _____

Draw 3 _____ % $ _____

Draw 4 _____ % $ _____

Draw 5 _____ % $ _____

Payments may be withheld on account of

1. Defective work not remedied,

2. Claims filed,

3. Failure of the Contractor to make payments properly to subcontractors or for labor, materials, or equipment,

4. Damage to the Owner or another contractor, or

5. Persistent failure to carry out the Work in accordance with the Contract Documents.

Final payment shall not be due until the Contractor has delivered to the Owner a complete release of all liens arising out of this Contract or receipts in full covering all labor, materials and equipment for which a lien could be filed, or a bond satisfactory to the Owner indemnifying him against any lien. If any lien remains unsatisfied after all payments are made, the Contractor shall refund to the Owner all moneys the Owner may be compelled to pay in discharging such lien, including all costs and reasonable attorneys' fees. Owner may withhold a retainage of 20% of all invoiced charges if the Contractor fails to complete all contract items upon submission of final invoice.

This Agreement entered into as of the day and year first written above by:

OWNER(S)

_____ (signature)

_____ (signature)

CONTRACTOR:

_____ (authorized signature)

Company Name and Address Goes Here

Items in a standard contract (above) are written to protect the buyer rather than the contractor. To protect your own interests, make sure the items listed in this chapter are included in an addendum that's attached to and becomes part of the signed contract.

• A list of what is not included (see Protecting the Floor During Construction on page 26)
• Provisions for obtaining written approval from the client once a final sample has been agreed upon and before the work starts
• A discussion of the nature of overlays and toppings, reinforcing that these materials are not prefabricated and will not be totally uniform or predictable
• Payment schedule
• The need for unrestricted access to the project during surface preparation, installation, and sealing
• Curing times and how long the work needs to be protected
• Special safety provisions, such as proper ventilation and elimination of all possible sources of open flames or sparks when the job requires solvent-based sealers and dyes (read Chapter 6 for a list of additional safety precautions)
• Start and completion dates of the project marking the total duration of the entire installation process, from preparing the surface through final sealing
• Labor rates for standard work hours as well as overtime or weekend work
• How much advance notice is required prior to starting the project
• Liability responsibilities (general liability insurance, bonding, workers' compensation insurance, etc.)
• Stipulations for cancellation, including a designated grace period for cancellation by either party
• Site issues such as accessibility, the need for the building to be dried in, barricading of the finished work, and sufficient lighting, power and water
• A disclaimer stating that even a well-executed overlay installation, complete with the necessary repair of existing flaws in the substrate, could still develop cracks

Regarding scheduling, don't set yourself up for failure. Build into the schedule a buffer of a day or two in case work is delayed by rain or other unexpected events beyond your control. Barring such events, however, if you say you're going to show up to perform your work on a specific date, be there. Delaying the project without good cause will make you look unprofessional. If for whatever reason you must make a change in the schedule, call your clients to inform them of the change as a common courtesy.

Fair contracts are a balancing act: The owner must be given all the important details yet not be scared away by a multipage laundry list of possible problems.

Having a fair contract also means being flexible when the unexpected occurs. A true professional will make every effort to work with schedule changes or other obstacles that may crop up.

Cutting decorative score lines in concrete can generate a lot of dust. Be sure to wear a face mask and safety glasses to protect your eyes and lungs from irritation.

When spray applying solvent-based dyes, always wear a face mask or respirator to prevent inhalation of hazardous fumes.

CHAPTER 6

THE IMPORTANCE OF SAFETY

It's all too easy to get caught up in the creativity of working with decorative toppings and gloss over safety. Don't do it. Many contractors start a project with the end in mind (a beautiful piece of work), but fail to take into account everything it takes to attain that goal, starting with safe working conditions. Not only can a disregard of safety procedures be harmful to you and your employees, you may also be putting other trades, building occupants, and even the environment at risk. Such negligence can be extremely costly, resulting in hefty fines from the Occupational Safety and Health Administration (OSHA) or the Environmental Protection Agency (EPA).

Safety starts with establishing a comprehensive jobsite safety program and putting someone in charge of overseeing it. This person should keep abreast of all safety rules and regulations, conduct regular safety talks with employees, monitor safety compliance, and make sure the appropriate Material Safety Data Sheets (MSDS) are on site. Per OSHA regulations, it's mandatory for contractors to conduct weekly safety training meetings.

Many contractors also conduct jobsite safety training to reinforce the lessons learned.

When I'm out on a jobsite, I certainly focus on the work at hand. But I also pay attention to what's happening around me, watching for situations that can spell disaster, such as someone smoking while working with a flammable substance.

It's imperative that you or a designated safety person scans the entire project site at all times!

The list of jobsite safety considerations is so extensive that I can't possibly cover all of them in detail in this guide. However, here are recommendations for handling potential hazards you may encounter when applying decorative toppings or overlays:

• Take precautions when working with cement-based products. Cement is very caustic and can cause severe skin burns, even after brief contact. Always protect your hands with nonabsorbent gloves. Also wear safety glasses to keep splatter from topping materials out of your eyes, especially during the mixing process.

• Take precautions when working with solvents. Many sealers are solvent-based, and you may also need to use solvents to clean application tools. Solvent-based products are extremely flammable and can be hazardous to breathe. Keep all possible sources of ignition away from the work area, such as lit cigarettes, open flames from a space heater, or sparks from a power saw or other equipment. If you're working indoors, make sure the area is well ventilated. Always wear eye protection, solvent-resistant gloves, and a respirator designed to protect you from hazardous fumes.

• If you plan to acid etch the concrete surface, you should always wear eye

Even small engraving tools can send tiny chips of concrete flying. Always wear the appropriate eye and face protection.

protection, nonabsorbent gloves, and protective clothing. Like cement, acid solutions are very caustic and can cause eye and skin burns.

• When working with cement-based toppings or powdered release agents (for stamped overlays), the powdered material can become airborne and present a breathing hazard. When mixing or applying these powdered materials, workers must wear a respirator or dust mask to protect their lungs, per OSHA regulations.

• Protect your knees from wear and tear by wearing kneepads whenever you are kneeling on a hard concrete surface. This may be one of the most overlooked safety considerations in our industry, and the consequences can be disabling. Just ask some of my co-workers and friends who have had to undergo surgery because they worked for prolonged periods on their knees without wearing adequate knee protection. (For overlays requiring you to be elevated during finishing, wear kneepads while working off spiked kneeboards.)

• Bags or buckets of overlay materials can be heavy, sometimes exceeding 60 pounds. Learn to lift properly (bending at the knees and lifting with your legs) so you don't throw out your back. Wearing a back belt for support can also help protect you from improper lifting.

• When using power equipment, always follow the recommended safety procedures for operation and wear the proper safety gear, such as safety glasses, work gloves, and hearing protection. As with kneepads, the use of proper hearing protection is often ignored. Serious hearing impairment can occur from prolonged use of loud power equipment, such as grinders and cutoff saws. I admit I have been guilty in the past of not protecting my ears with plugs or earmuffs—until I started noticing it was affecting my hearing.

• Check extension cords to make sure they are properly grounded and have no cut or frayed portions. Never use electrical equipment around water.

• Provide safe access when delivering raw materials to the jobsite or delivering mixed materials from the mixing station to the application site. Make sure no obstacles are in the way. I have seen instances where a worker slipped or fell while carting material in a wheelbarrow or buggy because the floor was slippery or a stray object on the floor became a tripping hazard.

• Many overlay systems require the applicator to wear spiked or cleated shoes. If you have never worn spiked shoes, they may take some getting used to. They make walking unstable, so it's very easy to turn an ankle. Try lifting your feet straight up and then straight back down. And beware of slick surfaces. Concrete substrates can be especially slick after a primer has been applied.

• If you're finishing an overlay using a fresno with a long aluminum handle, be careful of any overhead electrical wires. You could be electrocuted if the metal contacts exposed wires. A good way to avoid this problem is to use a fiberglass handle.

• When using larger equipment, such as a mortar mixer or pump, make sure the wheels are chocked so the machinery doesn't roll or move. Also, when operating gas-powered equipment such as a generator, make sure you have plenty of air movement and cross ventilation, especially when working indoors.

• Don't ignore environmental safety. Capture and dispose of any hazardous residue left behind from your work. This is getting to be a very serious problem in some parts of the country. Protecting the environment should be a crucial aspect of any jobsite safety program.

Safety Resources on the Web

American Society of Concrete Contractors' safety resources:
www.ascconline.org

American Society of Safety Engineers:
www.asse.org

Environmental Protection Agency:
www.epa.gov

MSDS online (a searchable database of MSDS documents):
www.msdsonline.com

National Institute for Occupational Safety and Health:
www.cdc.gov/niosh

OSHA: www.osha.gov

OSHA in Spanish:
www.osha.gov/as/opa/spanish/index.html

SafetyInfo (a library of online safety resources):
www.safetyinfo.com

Safety is not a game. You can only protect the well-being of yourself and your employees by implementing a safety program and creating a safe work environment. While working on a project, don't just focus on a specific task. Be aware of your surroundings and always be on the lookout for potential hazards.

MARSHALLTOWN CO.

Don't neglect to wear kneepads whenever you are kneeling on a hard concrete surface for prolonged periods.

The condition of the existing substrate is a major factor in the success or failure of an overlay. The concrete surface must be clean and free of any scaling, grease, dirt, or other grime that could inhibit bonding of the topping.

CHAPTER 7

SITE CONDITIONS AFFECTING THE INSTALLATION OF DECORATIVE OVERLAYS

With such a variety of overlay systems to choose from, you will discover that many products have their own set of performance criteria—as well as specific recommendations as to the best conditions and procedures for application. Often these criteria will help you determine the best system to use for a particular project, as discussed in Chapter 1. But whatever system you choose, I can't stress enough the importance of reading the manufacturer's technical data sheets and specifications. This is where you'll find vital information about mix ratios and mixing times, the best temperatures for installation, surface preparation procedures, and much more (see Chapter 12).

Most important, the manufacturer's specifications will describe when (or when not) to install their system based on such factors as exposure conditions, subfloor type, the age and moisture content of the slab, and the condition and profile of the existing substrate. Do the necessary groundwork beforehand to find out what site conditions could affect the system you're installing. Often these issues are impossible to correct once the overlay goes down.

Exposure conditions

Whether you are working indoors or out, one of the most important considerations when installing overlays is temperature. Quite simply, the warmer the air temperature and the substrate temperature, the faster the material will set.

Installing an overlay in direct sunlight during the hottest part of the day will greatly reduce your working time and jeopardize the quality of the results. In the heat of summer, keep your raw materials cool and install the overlay mix in the morning or during the coolest part of the day to extend the working time.

In the cooler winter months, you may encounter the opposite problem and find that your topping sets too slowly. Check the technical data sheets for the overlay to determine optimum temperature ranges for installation. For interior jobs, it may be necessary to heat the rooms you are working in prior to overlay installation.

Air movement during overlay installation also needs to be controlled to avoid premature drying, which can cause the freshly placed overlay to skin over during the troweling stage. If you have experienced this on a job, you know how difficult the situation is to control. Sometimes the project schedule may require you to install a floor overlay before the building is completely enclosed. That means you'll need to block off windows or doorways with plastic or plywood to stop as much air movement as possible. If you're working outside, try to install the topping in the morning, when the winds are usually the calmest. If you are forced

On large commercial projects, you'll often be competing with other trades. If possible, arrange to have sole access to the floor while installing the overlay to avoid interference and potential damage to surface.

DECORATIVE CONCRETE INSTITUTE

Don't open the overlay to traffic before allowing it to cure completely.

Protect Your Work!

Don't let all your painstaking efforts go to waste by failing to protect your newly placed overlay from unwanted traffic—from passersby, homeowners, or other construction trades.

Someone walking or driving on your new installation can spell disaster. Check the technical data sheets for the recommended cure times before opening the overlay to traffic.

Depending on the size of the project, this could be several days or more. For long-term protection, be sure to apply a sealer or coating, as discussed in Chapter 21.

to install on a windy day, pour smaller sections if possible. You can also set up temporary wind breaks by erecting a wall of tarps or by strategically positioning your construction trucks in front of the work site.

Subfloor type

Inevitably, questions arise about the performance of overlays or underlayments over wood subfloors. Again, read the technical data sheets! Some manufacturers strongly warn against applying their products over wood subfloors while others say it's OK if certain precautions are taken. The main consideration when going over wood is the amount of deflection the floor exhibits, which is usually determined by the joist spacing and the thickness of plywood used. On wood floors requiring more rigidity, we have had good success screwing down 3/4 inch plywood on 12 to

16 inch centers. (More details about this approach are provided in Chapter 14 on underlayments.)

Condition of the existing substrate

As a general rule, most overlay manufacturers recommend curing new concrete a minimum of 28 days before applying their products. This waiting time permits some of the moisture to escape from the slab. If toppings are installed prematurely, moisture vapor passing through the capillaries of the concrete can build up underneath the topping and lift it. For some overlay systems, manufacturers will stipulate a maximum moisture vapor emission rate. Read Chapter 8 for more information about testing for moisture emissions.

Proper surface preparation is by far the most important factor in ensuring good bonding of the overlay. The concrete

Construction scaffolding, forklifts, and jobsite debris: These are just a few of the assault weapons other trades could use to wreak havoc on your newly placed overlay. Be sure to protect your work both during and after placement.

When installing an overlay outdoors, pay close attention to the weather forecast. Extreme temperatures, wind, direct sunlight, and rain can all jeopardize the quality of your results.

surface profile, or degree of roughness, must be appropriate for the product you're applying. The importance of obtaining the right surface profile and various surface preparation methods are described in Chapter 9. But let the manufacturer's instructions be your primary guide. Some manufacturers may say that a mild acid wash is an acceptable means of profiling your floor while others advocate shotblasting.

Protecting the existing surroundings is equally as important as protecting your work. Taping off adjacent surfaces with a tarp or plastic sheeting is especially critical when installing spray-applied toppings or colored stains and dyes.

CHAPTER 8

DEALING WITH MOISTURE-VAPOR TRANSMISSION

Improper surface preparation is by far the leading cause of topping and overlay failure. But another potential source of problems is moisture within the concrete. Some resurfacing systems are impermeable, which means they act like raincoats to shield the concrete from moisture penetration. But these same systems can also trap moisture inside the concrete, restricting the ability of the concrete to "breathe" or release moisture vapor as needed. The consequences: As the moisture vapor travels up through the capillaries of the concrete and reaches the surface, it can break the bond of the overlay system, causing delamination.

Moisture-vapor transmission vs. moisture content

The culprit isn't actually the moisture content of the concrete, but moisture movement. Many overlays and toppings can bond to a concrete slab with a high moisture content. Only when conditions are right to cause movement of that moisture to the surface do problems occur. Water vapor will migrate to the surface of the substrate and become trapped beneath the overlay when the vapor pressure is higher in the concrete than in the surrounding atmosphere, due to differences in temperature and humidity levels. The moisture content of unprotected concrete that has been in service for some time tends to equalize with that of the atmosphere. So you'll encounter problems more often in newly placed concrete or in a slab-on-grade without an adequate vapor barrier beneath it.

How to test for moisture vapor

Here are two simple ASTM tests you can conduct to help determine whether a concrete slab is emitting too much moisture vapor. I recommend performing these tests before installing any impermeable overlay or topping system on projects where excess moisture is a concern. Be sure to follow the directives of the overlay manufacturer regarding the maximum allowable substrate moisture content.

• *Plastic Sheet Method (ASTM D 4263)*. Place an 18x18-inch sheet of clear plastic over the concrete slab and seal all four sides with duct tape. Leave it in place for 16 hours, and then check for condensation on the underside of the plastic. If visible moisture is present, the slab is too wet for overlay application. This test will only reveal the presence of moisture, but not the quantity.

• *Calcium Chloride Test (ASTM F 1869)*. This test provides the greatest accuracy because it will tell you the actual amount of moisture being transmitted by the concrete surface. Purchase a moisture-vapor test kit, which includes a small container of preweighed, unhydrated calcium chloride. Remove the lid, and place the container on the concrete under a sealed dome for a period of 60 to 72 hours. Then weigh the container to determine how much moisture left the concrete and was absorbed by the calcium chloride (the rate of moisture absorption is calculated in pounds of water per 1,000 square feet of surface area). Compare the measured rate to the acceptable maximum value provided by the overlay or coating manufacturer.

Note: Moisture-vapor test kit results will only be useful in predicting potential problems when the tests are conducted under the same environmental conditions the concrete will be exposed to while in service. During the test period, measure the air temperature and relative humidity levels to make sure they are within the anticipated ranges.

Ways to control moisture transmission

The best way to eliminate or reduce moisture-vapor problems is to prevent them in the first place by following good construction practices during concrete placement. Preventive measures include placing the concrete directly over a puncture-resistant vapor barrier, using concrete with a low water-cement ratio, and adequate curing of the slab.

But what happens if you're asked to install a resurfacing system over an existing slab that tests high for moisture-vapor emissions?

Here are some solutions:
• If the concrete is newly placed, allow it to cure for at least 28 days to give excess moisture time to evaporate.

• If possible, use an overlay or topping that permits free passage of moisture, rather than an impermeable system that acts as a vapor barrier.

• Use a moisture-control system, a surface-applied product designed to penetrate the pores of new or existing concrete to decrease permeability and suppress moisture-vapor movement. A number of these systems are available. Consult with your overlay supplier for recommendations.

A calcium chloride test will tell you the actual amount of moisture being emitted by the concrete, allowing you to compare the measured rate to the acceptable value provided by the overlay manufacturer. Test kits are available that include preweighed containers of unhydrated calcium chloride.

A simple way to test for too much moisture vapor in a slab is to tape plastic sheeting to the concrete surface and then wait for at least 16 hours. If you see evaporation accumulate on the underside of the plastic, the moisture content of the slab is too high for overlay application.

CHAPTER 9

SURFACE PREPARATION: OBTAINING THE RIGHT PROFILE

If you don't do a sufficient job of preparing the substrate for a decorative overlay, all of your painstaking artistry could be a waste of time and effort. I would bet that 90% of overlay delaminations and failures are due to inadequate surface preparation.

This critical step involves more than simply cleaning the substrate and removing existing coatings. Obtaining the proper concrete surface profile, or CSP, is equally important (see What's a 'CSP' Number? on page 42). There are several methods you can use to profile the concrete surface, including chemical means such as acid etching or mechanical methods such as shotblasting, diamond grinding, scarifying, and sandblasting. Be sure to follow the overlay manufacturer's recommendations for the best method to use, since the requirements may vary depending on the properties of the system you're applying.

Mechanical profiling

The best way to remove most contaminants and unsound concrete is by mechanical methods. Mechanical abrasion not only breaks up existing sealers, coatings, or adhesives, it will also lightly pulverize the surface, leaving a roughened profile for overlays and toppings to grab on to.

Shotblasting is often one of the most cost-effective methods of removing contaminants from a large area and for prepping substrates for self-leveling or polymer overlays, according to the International Concrete Repair Institute (ICRI).

Shotblasters use centrifugal force to propel steel shot at high velocity onto the surface. The big advantage of shotblasting when compared with other mechanical abrasion methods is that it produces very little airborne dust or debris. The process is confined in an enclosed blast chamber that recovers and separates the dust and reusable shot.

The depth of removal is controlled by a combination of factors, including the shot size and rate of machine travel. ICRI recommends a maximum removal depth of 1/4 inch per pass. But for most overlay systems, that much removal won't be necessary unless you need to strip away thick coatings or adhesives. If that's the case, you may need to make multiple passes with the shotblaster or consider a more aggressive removal method, such as a scarifier.

Dustless grinding is another effective way to prepare a concrete substrate for an overlay. Many of today's dustless grinders can serve multiple functions. For profiling tasks, they can be fitted with diamond-segmented abrasives of various grit levels ranging from fine to coarse, depending on how aggressive the profiling needs to be.

It's not uncommon to encounter floor surfaces that are buried under heavy adhesives or tile mastic. Removing the gook won't be easy, but it can be done using aggressive chemical strippers or a scarifier.

The International Concrete Repair Institute offers actual samples of the different concrete surface profiles to give you visual and tactile examples of the varying degrees of roughness.

A grinder equipped with the proper tooling can strip away even thick coatings and adhesives.

The Dust Factor

Note that some mechanical abrasion methods (particularly those without dust-collection systems) can leave behind a layer of concrete dust, spent abrasives, or other residue that could affect the adhesion of your topping. Be sure to sweep, vacuum, or pressure wash the surface until it's completely clean and free of any unwanted surface contaminants.

41

What's a 'CSP' Number?

Overlays bond best to surfaces that have a rough, sandpaper-like finish. But how do you know when you've achieved the right degree of roughness?

To help contractors make this assessment, the International Concrete Repair Institute has developed benchmark guidelines for CSP (concrete surface profile)—a measure of the average distance from the peaks of the surface to the valleys. They range from CSP 1 (nearly flat) to CSP 9 (very rough). As a general rule, the thicker the overlay or topping, the more aggressive the profile needs to be. A skim coat, for example, may require a light CSP of 2 to 4. For self-leveling overlays up to 1/8 inch thick, acceptable profiles generally range from CSP 4 to 6. For thicker polymer overlays (up to 1/4 inch), a CSP of 5 or greater is usually recommended (see method selector graph on page 47).

Achieving surface profiles in the higher ranges often requires roughening by shotblasting, abrasive blasting, or scarifying.

For a copy of ICRI's technical guide, "Selecting and Specifying Concrete Surface Preparation for Sealers, Coatings, and Polymer Overlays," call 847-827-0830 or visit www.icri.org.

Caution! The texture and appearance of the profile obtained will vary depending on strength, the size and type of aggregate, and finish of the concrete surface. On sound substrates the range of variation can be sufficiently controlled to closely resemble the referenced CSP standard. As the depth of removal increases, the profile of the prepared substrate will be increasingly dominated by the coarse aggregate.

Images generated using video density imaging techniques are courtesy of David Lange, Department of Civil Engineering, University of Illinois at Urbana-Champaign.

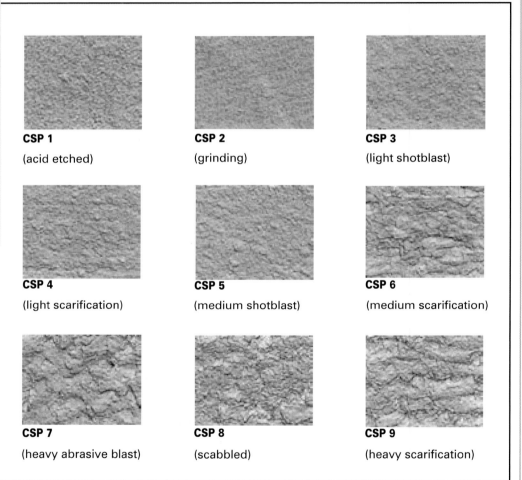

CSP 1 (acid etched)

CSP 2 (grinding)

CSP 3 (light shotblast)

CSP 4 (light scarification)

CSP 5 (medium shotblast)

CSP 6 (medium scarification)

CSP 7 (heavy abrasive blast)

CSP 8 (scabbled)

CSP 9 (heavy scarification)

Scarifying attachments can also be affixed to these machines for removal of thick coatings and mastics. Many contractors also use grinders with finer-grit abrasives to give traditional concrete surfaces and some overlays a high-polish luster. A dust-containment system vacuums up virtually all of the dust generated during grinding.

Acid etching

Some overlay manufacturers' specifications may approve of acid etching to profile the concrete substrate. This method typically involves applying a solution of water and muriatic or citric acid to the concrete to chemically remove the cement paste at the surface and lightly expose the fine aggregate.

However, there are definite limitations and drawbacks to acid etching that could make the method unsuitable for certain applications and environments.

Because acid etching produces a very light profile (similar to fine sandpaper), it's generally suitable only for skim coats or microtoppings less than 10 mils thick,

DECORATIVE CONCRETE INSTITUTE

A dustless grinder preps a floor for a cement-based topping. These versatile machines can be fitted with diamond-segmented abrasives of various grit levels to match your profiling requirements. They are also great for indoor use because a vacuum picks up nearly all the grinding dust.

What's the Best Dilution Ratio?

The surface hardness, or density, of the concrete will generally dictate the dilution ratio for the acid solution you use. For most jobs, the ratio will range from five to eight parts water to one part muriatic acid. On soft concrete, you may get by with a higher ratio of water to acid while on harder concrete you need to be more aggressive and reduce the dilution ratio to one to three parts water to one part acid. When mixing the acid solution, add the acid to the water instead of the reverse to prevent splashing.

Surface Prep Checklist

✓ Check the age of the concrete. New concrete slabs should cure for at least 28 days before overlay application.

✓ Remove all substances that could inhibit the ability of the overlay or topping to bond to the concrete. These include curing compounds, sealers, paints, and coatings.

✓ Remove any unsound concrete down to solid, clean concrete to provide a sound base for the overlay. These include areas of spalling, scaling, delamination, crumbling, or laitance (a thin layer of weak material containing cement and fines). To detect delamination, hit the surface with a hammer or other heavy instrument and listen for a hollow sound.

✓ Repair cracks and replace unsound concrete (see Chapter 10 for repair procedures).

✓ Remove all dirt, oil, or grease using a heavy-duty cleaner/ degreaser or trisodium phosphate wash.

✓ Test for moisture-vapor transmission levels using the plastic sheet test (ASTM D 4263) or calcium chloride test (ASTM F 1869), as described in Chapter 8.

✓ Give the surface the correct concrete surface profile for proper bonding of the overlay (see What's a 'CSP' Number? on page 42).

✓ After profiling, vacuum up any remaining dust or debris to leave a clean, dust-free surface ready to accept the overlay or topping.

Don't neglect to profile the edges of your floor. If you're working next to a wall, special grinding attachments are available for maneuvering right along the edge.

After obtaining the desired surface profile through mechanical abrasion, vacuum up any remaining dust and other residue to leave an immaculate surface for overlay application.

according to ICRI. You also need to take great care to neutralize the acid and remove all residual solution from the surface. Not only does the acid residue act as a bond breaker, it can also penetrate into porous concrete and chemically react with the cement, affecting the integrity of the concrete surface.

Cleaning up the spent acid is often a messy process requiring a lot of rinsing and scrubbing and copious amounts of water. Be sure to check your local environmental regulations regarding containment and disposal of the acid and rinse water. During the acid wash, contaminants lifted from the surface may contain toxic materials such as tile mastics, lead-based paints, or asbestos.

Worker safety is also a big concern when applying acid solutions. Because the acid is highly caustic, applicators must wear the necessary protective gear to prevent eye and skin burns. The acid could also react with and corrode any metal components it comes in contact with, such as electronic equipment and machinery, so these elements must be protected as well.

Before applying the acid solution, dampen the substrate so you don't get aggressive "burn" marks where the acid first hits the surface. Then apply the solution uniformly with a low-pressure sprayer (use one that's all plastic, with no metal parts) or a plastic sprinkling can. Follow immediately by scrubbing the acid solution with a stiff-bristle broom to work it into the surface. The acid will begin to bubble as it eats away at the surface paste. If you see areas where bubbling does not occur, you may need to apply additional acid solution.

When the acid stops bubbling (usually after 5 to 10 minutes), you must clean the floor thoroughly to remove all etching residue. I use a cleaning solution of 1 quart ammonia to 5 gallons water because this helps to neutralize the acid. A solution of baking soda and water is also an effective neutralizer. Scrub the slab thoroughly with the neutralizing solution and then rinse well with water. Remove the remaining residue with an industrial wet/dry vac or squeegee vac. I recommend a final power washing to ensure a completely clean substrate.

Without a doubt, surface preparation can be a tedious, time consuming, and messy task. But don't be tempted to take short cuts! Only by doing the job properly and thoroughly can you ensure long-term bonding of the overlay—and a satisfied customer.

Using a shotblaster is one of the best methods for preparing concrete substrates. The machine propels steel shot at the surface to blast away most contaminants and unsound concrete while leaving a roughened profile for overlays and toppings to grab on to.

Method selector

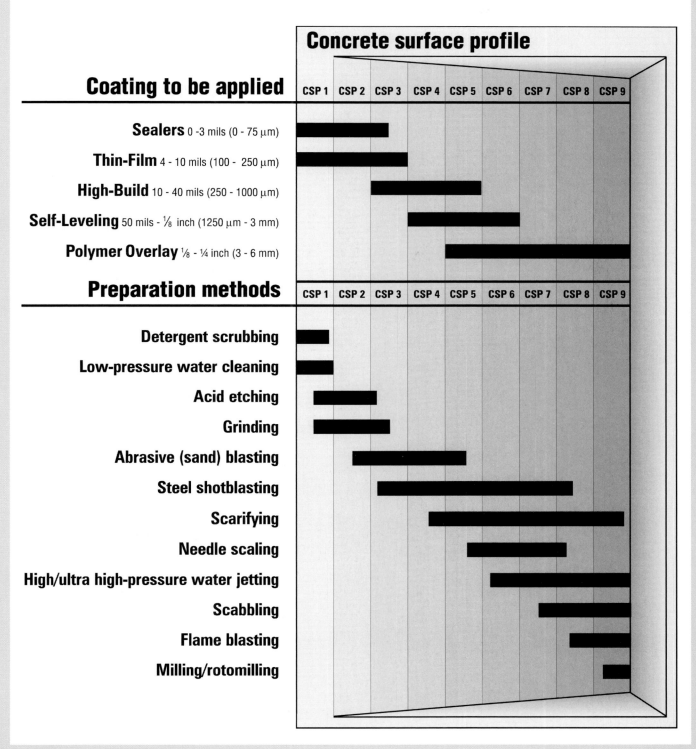

Concrete surface profile

Coating to be applied

	CSP 1	CSP 2	CSP 3	CSP 4	CSP 5	CSP 6	CSP 7	CSP 8	CSP 9
Sealers 0 -3 mils (0 - 75 μm)	■	■	■						
Thin-Film 4 - 10 mils (100 - 250 μm)	■	■	■						
High-Build 10 - 40 mils (250 - 1000 μm)			■	■	■				
Self-Leveling 50 mils - ⅛ inch (1250 μm - 3 mm)				■	■	■			
Polymer Overlay ⅛ - ¼ inch (3 - 6 mm)					■	■	■	■	■

Preparation methods

	CSP 1	CSP 2	CSP 3	CSP 4	CSP 5	CSP 6	CSP 7	CSP 8	CSP 9
Detergent scrubbing	■								
Low-pressure water cleaning	■								
Acid etching	■	■							
Grinding	■	■							
Abrasive (sand) blasting			■	■	■				
Steel shotblasting			■	■	■	■	■		
Scarifying				■	■	■	■	■	
Needle scaling					■	■	■		
High/ultra high-pressure water jetting						■	■	■	
Scabbling							■	■	
Flame blasting								■	
Milling/rotomilling									■

A colored primer is applied over repaired cracks in preparation for a self-leveling topping.
If you are meticulous with your crack repair efforts, it's unlikely anyone will know the cracks ever existed.

How do you know when a crack requires repair? If it's equal to or wider than the width of a credit card, get out the crack filler.

Even when you use the most sophisticated materials and cutting-edge techniques to fix cracks, there are no guarantees that the cracks won't re-form and mirror through to the overlay. When the existing concrete slab is veined with large cracks like these, tell your client that you can't promise success.

In some cases, you can use the crack to your advantage and incorporate it into the pattern of the overlay. To fill the existing expansion joints, use a flexible filler to allow it to expand and contract rather than trying to restrain it.

CHAPTER 10

REPAIRING CRACKS

Unfortunately, most concrete has a tendency to crack, usually as a result of improper design or poor construction practices. As the overlay installer, you have no control over these factors. But you are responsible for repairing any cracks in the existing concrete before resurfacing. Understanding the types of cracks and their causes will help you choose the best fix for the job, so your client will get a new surface that will last for years to come.

Keep in mind, however, that no method of crack repair will ensure success.

Despite your best efforts, the crack may eventually re-form and mirror through to the topping. But if you did the repair properly, any crack that does reappear is likely to be minor. Be sure to spell out in your contract that you will take all the steps necessary to repair cracks, but can't guarantee the results. The essentials of a good contract are discussed in Chapter 5.

Types of cracks

Cracks in concrete fall into two general categories: static and moving. Static cracks are hairline flaws that only affect the concrete surface. In most cases, they require little or no repair (see Cracks You Don't Have to Fix). Moving cracks, also called active cracks, are more serious. These cracks often are structural in nature and continue through the entire depth of the concrete. There are several causes, including insufficient spacing and sequencing of control joints, not isolating new concrete from old, and improper subgrade compaction.

If you encounter moving cracks, you typically must repair them before you resurface the slab. In some cases, however, you may be able to incorporate the cracks into the overlay design, as described later in this chapter.

What if you're unsure whether the crack is static or moving? Here's a rule of thumb used by crack repair specialists: If the crack is equal to or wider than the width of a credit card, it's probably a moving crack and needs to be repaired. For cracks that are borderline (not quite credit card width but slightly wider than hairline), consider taking the time to treat them as structural cracks for insurance.

Choosing a repair method

If you determine that the cracks in the concrete must be repaired before you can install the overlay, your next step is to choose a repair method. But first, you must do some research to find out the best product and process for your project and the overlay system you're using. Check with your material supplier for recommendations on the method of repair best suited for use with their product.

Following are some of the techniques I have used successfully to repair moving or active cracks. Regardless of the method you choose, it is essential that each crack be cleaned thoroughly prior to filling.

1. Crack chase, clean, and fill. Chase, or route out, the crack using an angle grinder with a V-grooved diamond blade. Remove all debris with compressed air or a shop vac and then fill the crack, preferably with a semi-rigid material or per the manufacturer's recommendations. Generally, I do not like to use flexible fillers because they exhibit too much expansion and contraction. You can't expect a more

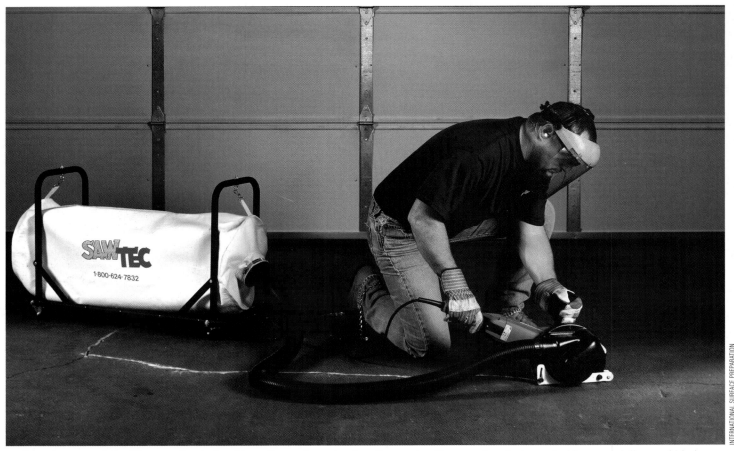

Prepping a crack for filler usually starts with routing out, or chasing, the crack with an angle grinder and a V-grooved diamond blade.

rigid cement-based product (your overlay) to adhere to and stretch along with a flexible filler. I have had great success using polyurea-based products because they are rigid but have some flex. Semi-rigid materials allow for slight movement but usually not enough for a crack to reappear in the overlay. In addition, these materials dry very quickly, allowing you to get back on the surface sooner.

On exterior concrete, some installers will chase the crack and then fill it with hydraulic cement, also called a "water plug," to prevent moisture from coming up through the crack. Be aware, however, that if there is subsequent movement of the crack, this cement-based product may break out over time. A water plug material works best with non-moving cracks.

2. Install new control joints. After filling cracks using the method above, consider saw cutting a new control joint close to the crack to help relieve expansion and contraction stresses in the slab. I strongly recommend using this procedure with structural crack repairs.

3. Epoxy injection. Epoxy injection is an excellent method of repairing structural cracks, but it can be time consuming and costly and requires great skill. This procedure consists of mounting ports in the crack surface at intervals, with the spacing approximately equal to the thickness of the concrete. Epoxy is then injected through the ports using a pump operating at a low pressure to ensure complete filling of the crack. Unlike the other methods described here, epoxy injection provides a structural weld when executed properly. That means the crack is unlikely to reappear. According to ACI 224.1R, cracks as narrow as 0.002 inch can successfully be repaired by injection with epoxy (see Resources for More Information on next page).

4. Crack stitching. Stitching a crack requires drilling holes on both sides of the crack and then spanning the affected area with wires or U-shaped metal strips, which are grouted or epoxied into place. Some installers saw cut a wide kerf across the crack and then epoxy rebar into the area to fuse the two sides together. As with epoxy

injection, this is a very effective method of repairing serious cracks but is time consuming and expensive. Also, you must place a sufficient amount of overlay material over the repair so it doesn't reflect through to the finished surface.

5. Use a repair kit. Some manufacturers have crack repair kits that include all the materials you need to make repairs before applying their overlay systems.

One advantage of using these kits is that you know the repair materials will be compatible with the resurfacing material. Another plus is that a kit will include all the essentials you need for an effective repair.

Making the crack part of the design

On many jobs, I have successfully incorporated the random pattern of the cracks into the design of the overlay. This is a great way to hide the crack aesthetically while making it functional. Here are the basic procedures:

• Route out the cracks with a

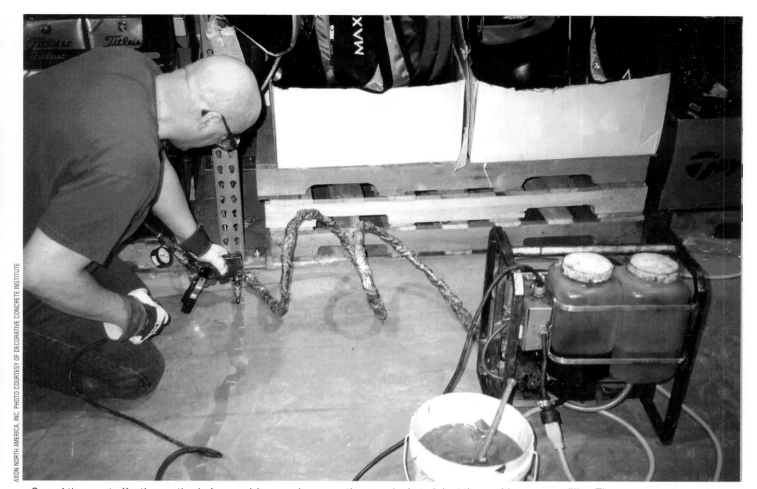

One of the most effective methods for repairing moving, or active, cracks is to inject them with an epoxy filler. The epoxy creates a strong structural weld that will keep the crack from re-forming.

V-shaped diamond blade.
- Apply the cementitious overlay material, letting some of it fill the routed cracks.
- Once the topping has reached sufficient strength (wait a minimum of one day), re-chase the crack.
- Fill the cracks with a flexible or semi-rigid filler to allow the crack to expand and contract naturally rather than trying to restrain it.

As part of the design, you can saw cut additional "faux cracks" where necessary, to mimic natural flagstone for example.

What to do about gapping cracks

In most cases when cracks are extremely wide and unstable, I recommend complete removal and replacement of the existing concrete. On one project, however, the client insisted on repairing the cracks and installing the overlay despite my reservations. This called for extreme measures, and my crew and I came up with a fix that worked.

In this particular slab, the gapping cracks were so unstable you could easily chip down through the entire depth of the concrete and expose the subgrade. We chased all of the cracks and then filled them with a water plug to stop the migration of moisture. Next, we attached expanded-metal lath to the entire surface of the concrete floor using a pneumatic nail gun. To encapsulate the metal lath, we then applied a 3/8-inch-thick coat of a self-leveling cementitious overlay. After allowing the overlay to dry, we came back and topped it with a 1/4-inch finish layer to provide a smooth base for chemical stain application. A year later, we returned to the project to see how well the floor was holding up. Fortunately, there were no hollow spots or cracks to be found.

As I mentioned previously, you should never attach a warranty to crack repair. In this case, we did inform the client before proceeding that the slab was in such bad condition that it was probably not a good candidate for an overlay.

Resources for More Information

The following publications offer more detailed instructions for repairing cracks in concrete slabs. Both are available from the International Concrete Repair Institute (www.icri.org).

ACI 224.1R, "Causes, Evaluation, and Repair of Cracks in Concrete Structures," American Concrete Institute.

"Cracks in Concrete: Causes and Prevention," a compilation of 14 articles reprinted from Concrete Construction magazine.

CHAPTER 11

ESTIMATING MATERIALS

The suppliers of materials for overlays and toppings usually include coverage rates and yields in their technical data sheets, making it easier for you to estimate how many bags or buckets of the product you'll need for a particular job.

You might assume, then, that ordering the right amount of material would be a no-brainer as long as you accurately measure the size of your project. In reality, though, the math isn't quite that simple.

To accurately estimate material quantities, you must also compensate for factors such as spillage and waste and the thickness of your placement. Here are some simple formulas you can use to ensure that you order enough material and never have to suffer the gut-wrenching feeling of running short during an installation.

Measuring project size

The most common method of estimating materials is based on the square footage of the area to be installed. This is easily determined by multiplying the length of the slab by the width: A project measuring 30 x 30 feet, for example, would equal 900 square feet.

For projects with areas that aren't perfect squares or rectangles, the best strategy is to break down the areas into rectangular

sections, or grids, to determine the square footage. Where you are left with irregular or pie-shaped sections, find the halfway point and average in both directions. You can obtain surprisingly accurate estimates using this approach, especially for large areas.

Another option when calculating the dimensions of large surfaces is to use a measuring wheel (one that measures in linear feet).

Coverage rates for overlay materials are based on the installed thickness. So if the technical data sheet says that a bag (or one unit) of self-leveling overlay placed at a thickness of 1/4 inch will cover approximately 25 square feet of surface area, you would take the square footage of your project and divide it by 25 to calculate the number of bags, or units, needed:

900 sq. ft. ÷ 25 = 36 units of material

Some installers prefer figuring out the area in cubic feet and then dividing that figure by the cubic footage per unit of material. (The cubic foot coverage rate will need to be provided by the manufacturer and is sometimes given on the bucket or bag the material is packaged in.) Though a bit more complicated, since you have to figure out the number of cubic feet of coverage per unit, this is a more accurate estimating method if you plan to apply the material at a thickness different from the one given in the technical data sheet.

Let's return to the previous example, using a 900-square-foot surface and a coverage rate based on a 1/4 inch material thickness. In this case, however, you plan to place the overlay at a 5/8-inch thickness.

Test Your Math

Now that you've taken Estimating 101, it's time to practice what you've learned.

Let's say you're installing an 1/8-inch-thick microtopping. The coverage rate for one bag of material, applied at 1/8 inch, is approximately 80 square feet. The dimensions of the floor you're resurfacing are 45 feet wide by 55 feet long. How many bags of material will you need, including enough for overage?

Here's the formula:
45 x 55 = 2,475 sq. ft.
2,475 sq. ft. ÷ 80 = 30.93 units, or 31 bags

31 x 10% = 3.1 (round up to 4)

31 + 4 = 35 bags of material to cover the floor at an 1/8-inch thickness

Here's a formula to convert square footage into cubic footage:

30 x 30 feet = 900 sq. ft. at 5/8-inch thick
5 ÷ 8 = 0.625 inch ÷12 inches = 0.052 inch
0.052 inch x 900 sq. ft = 46.8 cubic feet
Divide 46.8 by cubic footage per unit

This formula may seem confusing at first, especially if you have never figured thin toppings in cubic feet. But once you run through the exercise a few times, you will find this to be a very accurate method of performing your material take-offs.

Factor in overage

Once you determine the number of packages or units of material required, add in 10% to 15% more material than what your original take-off called for. This overage will help to compensate for unanticipated material use due to:

- Imperfections in the subfloor that use up more material. Examples include low areas, undulations, chips, and gouges.
- Spillage of material during mixing and handling.
- Jobsite conditions that cause a batch of material to set too fast, forcing you to dispose of it.
- Material wasted during pumping on larger overlay projects, as you adjust the amount of water needed.

In our earlier example calling for 36 bags of material, the formula for calculating overage is: $36 \times 10\% (0.10) = 3.6$ units, or 4 additional bags.

As when measuring square footage, always round up to the nearest whole number.

Estimating other materials

On most projects, you will need to perform estimates for other materials used during the installation, including primer, sealer coats, dry-shake hardeners, and stains or dyes.

Generally, coverage rates for sealers, stains, and dyes are also calculated by square footage. For example, the technical data sheet for the sealer you plan to use says the product has a coverage rate of approximately 250 square feet per gallon. As discussed previously, take the overall square footage of your project and divide it by the coverage rate:

900 sq. ft. ÷ 250 = 3.6, or 4 gallons of sealer

If the sealer is available in 5-gallon pails, it might be cheaper in this case to purchase the 5-gallon container rather than four 1-gallon units (often larger containers are more economical). And it won't hurt to have some extra on hand in case you run into problems.

Using up leftovers

Estimating for overage is good insurance that you won't run out of material during a project, even if you encounter the unexpected. But it also means there will be jobs with some material left over at the end. Don't dispose of it. Use the excess to create sample boards to display in your showroom or to resurface a floor in your office. This is your opportunity to try out some of your creative ideas and hone your skills. Have fun and experiment!

The first step in estimating materials is to determine the square footage of the area to be covered by multiplying the slab length by the width. Use this figure to calculate the number of bags of material needed based on the coverage rate per bag at the installed thickness.

CHAPTER 12

FOLLOW THE TECHNICAL DATA SHEETS

In today's marketplace, you can choose from a host of different decorative overlay products. While it's great to have such a wide array of choices, you'll discover that each system is unique and may have different requirements for surface preparation, mixing, application, and curing.

For example, some manufacturers may require shotblasting for surface preparation, while others say acid etching is sufficient (review Chapter 9 for descriptions of the various surface preparation methods). Some systems require application of a primer, while others simply require the concrete to be "saturated surface dry," or SSD, which means the permeable voids are filled with water but no water is on the exposed surface.

That's why it's absolutely essential for you to read, in their entirety, the technical data sheets provided with every product you use. Manufacturers write these specifications for a reason, and it's crucial to adhere to their guidelines.

What the data sheet will tell you

You'll find an abundance of valuable information in the typical tech data sheet for an overlay, including:

- Coverage rates
- Product shelf life
- Recommended applications and product limitations
- Surface preparation guidelines
- Mix ratios and required mixing times
- Application procedures
- Coloring methods
- Cure times
- Suggested sealers
- Protection and maintenance of the finished floor
- Performance data (such as compressive strength and abrasion resistance)
- Precautions

To save time or a few pennies per square foot, installers often are tempted to waver from the manufacturer's suggestions on mix ratios, troweling times, or how to properly prepare the substrate. But it is imperative that you stick to the instructions given in

CONCRETE SOLUTIONS, INC.

Read through the technical data sheets provided with every overlay product you install. They will tell you essential details about the requirements for surface preparation, mixing, application, and curing. Don't be tempted to improvise. You risk voiding product warranties by not adhering to the manufacturer's guidelines.

the tech data sheets. And those guidelines should supersede the general advice given in this book. Even if you think your way is better, you risk voiding product warranties by not following the manufacturer's directives.

Beware of combining products

I'm a strong advocate of sticking with one manufacturer's product line once I've found a system that works well for a particular application. I won't, for example, mix one manufacturer's polymer with a different manufacturer's bagged mix and then apply another manufacturer's sealer. Whenever possible, I use a complete system.

That said, professional installers may, on occasion, combine systems from different material suppliers if one manufacturer doesn't offer a specific product they need

to achieve a certain effect. Before doing this, consult with a technical representative to find out about the compatibility of products from other sources.

At our facility, we go one step further. We conduct extensive testing when combining products from different manufacturers to check for compatibility and performance. We'll install a trial placement and put the surface under extreme conditions, such as driving a forklift over it or sliding pallets across it. Before you combine products from different suppliers on an actual project, I suggest you conduct similar performance tests to ensure you'll get the results you expect.

PHOTO GALLERY

EVERLAST CONCRETE. PHOTO COURTESY OF BOMANITE CORPORATION

DISTINCTIVE CONCRETE. PHOTO COURTESY OF COLORMAKER FLOORS

ELITE CRETE SYSTEMS, INC.

CONCRETE SOLUTIONS, INC.

RUDD COMPANY, INC.

COLORADO HARDSCAPES

CHAPTER 13

SETTING UP A MIXING STATION

As you learned in Chapter 9, successful installation of a topping begins with preparing the surface properly. However, of equal importance is proper setup of the station where you'll be mixing all your materials. Unfortunately, some installers don't pay enough attention to this important aspect of the job. Not only should the mixing station be outfitted with all the equipment and supplies you'll need to mix materials properly (read more about equipment requirements in Chapter 22), it should also provide efficient access to the point of placement so you can deliver materials in a timely fashion.

Here are other key issues to consider when setting up a mixing station.

Protect surrounding surfaces

Short of grinding or acid etching, it's nearly impossible to remove any spillage that occurs during mixing. So your first task is to protect the surface you are mixing on, especially if you're working indoors. Cover the entire area with a drop cloth, building paper, plastic, or similar protective covering. Also lay out a paper or plastic covering over the path running from the mixing station to the drop-off point. This will protect the path from any material that gets on the bottom of your shoes and on the wheels of the concrete buggy or wheelbarrow you're using to cart materials. And don't neglect to protect the floor you're resurfacing if you need to carry or cart materials across it. The last thing you want to do is contaminate the surface you worked so hard to profile, repair, and prime.

Horizontal surfaces aren't your only concern. Mixing can create splatter that gets onto nearby walls and other vertical surfaces,

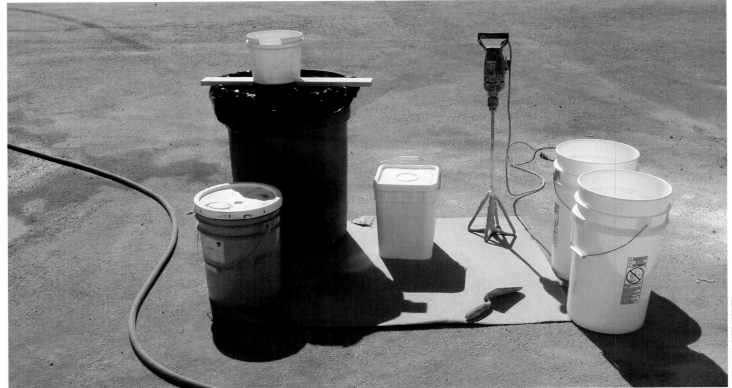

DECORATIVE CONCRETE INSTITUTE

Be ready for action by setting up your mixing station with all the equipment and containers required. You'll need a large vessel to hold the liquid component of the overlay mix, a separate 5-gallon bucket of water to dip tools into for cleaning, and a measuring bucket to ensure the addition of the right amount of material.

Easy access to the worksite is important when locating the mixing station, but don't set up your equipment too close to the slab being resurfaced. If you're mixing cement-based products outdoors, some of the dry material could drift onto the concrete surface, where it will act as a bond breaker.

Be sure to protect the floor you're resurfacing if you need to cart materials across it. Accidental spills won't be easy to remove.

especially if you're working in a confined space. Take the time to mask off these surfaces with plastic or paper sheeting.

Don't set up too close to the work area

While easy access is important when locating your mixing station, don't set up too close to the slab being resurfaced. If you're mixing cement-based products, some of the fine dry material could go airborne before it's thoroughly blended with the liquid component and end up settling on the concrete surface, where it will act as a bond breaker. In fact, whenever you're mixing outdoors, be alert to anything downwind of the mixing station. We learned that lesson on one job, when some of the material we were mixing went airborne and landed on a very expensive car. Fortunately, the car owner accepted our sincere apologies—and a car wash on us.

Address noise and odors

Is your resurfacing project in an occupied building or within shouting distance of a home or business? If so, be sensitive to the ears and noses of the people around you by minimizing the use of loud mixing equipment and products that emit odorous fumes, such as solvent-based materials. Proper ventilation is imperative when you're mixing materials in occupied indoor spaces or near food-preparation facilities, such as a restaurant.

Provide backup sources for power and water

You can't mix most topping materials without having power and water on hand, so it's essential to ensure an uninterrupted supply. Even on jobs where electrical outlets and working plumbing are readily available, you can't rely on these sources alone, as I've learned firsthand. We were making excellent

progress on a large mall construction project when a backhoe hit the water main and completely shut off our source of water. Luckily, we had the foresight to fill four 55-gallon drums with water, which allowed us to continue our mixing operation.

On another job, lightning struck the building we were working in and knocked out the power supply as we were trying to mix and place a skim coat. This time, our backup generator (with a full tank of gas) saved us.

Also be sure to check out the location and capacity of your power supply. On some jobs, you may need 220-volt three-phase power to operate certain equipment, such as a pump. Also have extension cords on hand in case you need to plug into a distant outlet.

Don't forget provisions for cleanup

After mixing, you'll need to clean your mixing tools and equipment promptly before the material sets. I keep 5-gallon buckets of water on hand to dip tools into for immediate cleaning.

Material disposal is another concern. Overlay materials can be expensive, so try to mix only the quantity you need. If you end up with leftover mixed material, make good use of it. For example, create sample boards by smoothing the mixture onto cement backerboards. You can use these samples to experiment with the effects of different stains, dyes, or sealers. I also use them to test my engraving tools, sawcutting equipment, and blades to see how effectively they score the topping.

If you still need to dispose of leftover mixed material, dump the excess into one of the empty bags or buckets the material came in. Then, once the material has set, dispose of it in a dumpster or large trash bag.

As you can see, setting up a mixing station involves much more than just grabbing a couple of buckets and dumping in your materials. By being prepared and covering all the bases, you can avoid a lot of potential problems.

DECORATIVE CONCRETE INSTITUTE

The mixing station should serve as the hub for blending all the materials needed for your project, including water-based pigments.

Tips for Measuring Materials

Many overlay systems require a very specific ratio of polymer or water to the dry components. But it can be very time consuming to measure out the required amount each time. Here are some timesaving tips.

• Let's say the product you're mixing requires 4-3/4 quarts of liquid per bag of dry material. Instead of measuring a quart at a time, take a plastic bucket and cut it down to the required 4-3/4 quart measure. Some manufacturers even provide a measuring bucket with each order to ensure the addition of the prescribed amount of material.

• Fill a vat or drum much larger than your measuring bucket with the liquid component you're using (either polymer or water). You can then easily dip your measuring bucket into the vat when you need to mix a new batch.

• Once you've filled the bucket, you'll need to set it on a level surface to obtain an accurate measure and to allow excess material on the sides to drip off. Put a piece of wood or a straightedge across the top of the vat you dipped the bucket into, and then rest the measuring bucket on it until the material levels off and stops dripping (it will all go back into the vat). This will reduce waste and save a lot of time and mess, especially on a large project.

69

CHAPTER 14

UNDERLAYMENTS

Underlayments, unlike their overlay cousins, do not get a chance to see the light. Their main purpose is as a thin topping to level, smooth, or fix imperfections in concrete substrates prior to installation of other floor coverings, such as tile, carpet, or vinyl.

Despite their undercover status, underlayments require the same care in installation as a self-leveling overlay (see Chapter 15 for details). Inadequate concrete surface preparation and priming might lead to severe pinholing or bubbling that could show through on some vinyl tile floors.

Installing over concrete

On new concrete subfloors, a minimum of 28 days curing is usually necessary before installing an underlayment. For most jobs, you can use the same tools required for overlay installation. You'll need a gauge rake to spread the material into place and to control the depth. Then you'll need a steel smoothing blade to level the surface and break the tension, allowing any entrapped air to escape.

You can place underlayments in thicknesses ranging from featheredge to 1-1/2 inches in one lift. It's even possible to achieve greater thicknesses where needed by extending the material with pea gravel (see Using Pea Gravel to Increase Thickness on page 71).

Installing over wood

It's also possible to install an underlayment over plywood subfloors, but additional steps are necessary to reduce the chance of cracking. Adding a liquid acrylic modifier to the underlayment mix will increase the flexibility and reduce the chance for cracking. Typically the modifier replaces a percentage of the water in the mix.

It's best to place the underlayment over 3/4-inch-thick plywood fastened with screws at 12-to 16-inch intervals to help decrease deflection and add stability.

On projects where deflection of the plywood is a concern, some manufacturers recommend applying the underlayment over two layers of 3/4 inch plywood for better rigidity. Once the plywood is properly secured, a sufficient amount of bonding primer needs to be applied to the wood substrate, primarily to seal off the wood.

When the primer has dried, attach galvanized metal lath to the plywood subfloor as a reinforcing material. I recommend screwing down the metal lath perpendicular to the direction of the plywood roughly on 6-inch centers, overlapping the sheets approximately 2 inches. Be sure to use enough fasteners to ensure flat, uniform installation. Some manufacturers suggest applying the underlayment directly over the top of the secured metal lath while others recommend first applying a repair-type mortar as a body coat over the lath. I like to use a polymer additive combined with cement, which provides some flexibility. However, make sure whatever product you use is approved by the underlayment manufacturer.

After the repair mortar cures (if using), install the underlayment to provide a smooth, level surface. When placing an underlayment over metal lath, be sure the thickness is at least 1/2 inch to reduce the chance of the lath pattern ghosting through to the surface. If you're going over an elevated wood subfloor that requires

When installed properly, an underlayment will correct unevenness and minor imperfections in concrete substrates to ensure a smooth base surface for tile, carpet, vinyl, and other floor coverings.

you to build thicker lifts, consider bringing in a structural engineer to make sure the floor will support the additional weight.

Keep in mind that even after performing all of the necessary prep work, it is common for cement-based toppings installed over wood to exhibit minor cracking. But in the case of an underlayment, this is usually not a concern since it will be covered up with some other flooring material.

Priming the substrate

Manufacturers may recommend priming the concrete or plywood subfloor to help seal it and reduce the chance for air vapor emitted from the substrate to form bubbles in the underlayment. Be sure to check the technical guidelines on primer recommendations, since some manufacturers require a different primer over wood than concrete.

If you're applying the underlayment to a wood subfloor in an area susceptible to moisture exposure, such as food prep facilities, it's a good idea to apply a flexible waterproofing membrane over the plywood to provide additional protection. A variety of waterproofing systems are available. I have had good success using one that consists of a mixture of pure acrylic and cement mixed at a one-to-one ratio. After mixing, I trowel the material onto the substrate and embed a fabric mesh in it.

Just because an underlayment will be hidden under a floor covering doesn't mean you can overlook surface preparation. Before installing the underlayment, be sure to profile and prime the floor and fill in any chips, gouges, or spalls with a suitable repair material.

An underlayment flows onto floor surfaces like thick pea soup and requires minimal effort to smooth and level.

You can place underlayments in thicknesses ranging from featheredge to 1 1/2 inches in one lift. Using a gauge rake to spread the material into place will help to control the depth.

CHAPTER 15

SELF-LEVELING OVERLAYS

Self-leveling overlays have been around for decades, but they were initially used for purely utilitarian purposes. In the late 1980s, these versatile overlays started coming of age as a legitimate floor finish alternative to pricier high-end materials such as slate, granite, and marble. Not only are self-leveling overlays less expensive than these alternatives, they may even outlast them when properly sealed and maintained. (See Chapter 21 for recommended sealing methods.)

Self-leveling overlays offer numerous advantages, for both the installer and end user:
• They provide a flat, durable surface that can serve as a blank canvas for a wide variety of decorative embellishments.
• They can smooth, level, and restore worn or uneven concrete. Because these overlays are typically applied at thicknesses of 1/4 inch or greater, they effectively cover minor flaws and compensate for height variances.

After pouring the freshly mixed overlay, you have a very limited working time. To avoid cold joints, establish predetermined starting and stopping points using divider strips or adhesive-backed foam weather stripping.

• They offer quick installation and rapid drying times. Some systems are ready for light foot traffic 4 to 6 hours after installation.

There are a few limitations, however. Because of their application thickness and the nature of the ingredients used, self-leveling overlays are less flexible after drying than microtoppings and spray-down systems, which can make them more prone to cracking. They also are prone to skinning over with any air movement across the surface, which jeopardizes their workability. For these reasons, self-leveling systems are generally restricted to interior applications.

Preparing the surface

Successful installation of a self-leveling overlay starts with proper surface preparation:

• If there are existing cracks, assess whether they are static vs. working or

Divider strips not only function as termination points for overlay placements, they can also serve as decorative enhancements. Strips of stainless steel, brass or wood are a few of the options.

In addition to the gauge rake, other application tools you'll need include a smoothing paddle to break the surface tension of the overlay and remove any entrapped air and a hand trowel to work in tight spots and to make a smooth transition when tying the pour into an existing placement.

Tip: Wear the Right Footwear

When applying the overly, wear cleated shoes with non-metal spikes, such as soccer shoes.

Metal spikes can puncture the primed surface. This allows air to escape from the substrate, leaving bubbles in the overlay surface.

moving. This will help to determine the best products and methods of crack repair to use. (See Chapter 10 for more information on repairing cracks.) Some overlay manufacturers offer patching products specifically for use with their systems.

• Most overlay manufacturers recommend honoring all contraction, construction, and isolation joints in the concrete substrate, which means you will need to mark their locations (such as on an adjacent wall) so that

later, when the overlay dries, you will be able to make sawcuts over the top of the existing joints.

• On newly placed concrete, self-leveling overlays require a minimum of 28 days curing time prior to installation. Conduct moisture-vapor tests (as described in Chapter 8) to ensure that the emission rate is not excessive. If the test results indicate that the rate is too high, you may need to cover the floor with a surface-applied moisture control product prior to overlay installation to bring the emission rate within industry tolerances. Check with the manufacturer for recommendations.

• Shotblasting is the preferred method for profiling concrete floors in preparation for self-leveling overlays. According to International Concrete Repair Institute guidelines, you should aim for a concrete surface profile of 4 to 6 (see Chapter 9).

Applying primer

After shotblasting the surface to obtain the desired profile, you must apply a primer. Don't neglect this important step. With self-leveling overlays, many failures and imperfections such as delamination, pinholing, and air bubbles are due to improper priming.

Priming serves two important functions. First, it seals the pores of the concrete so it won't absorb moisture from the newly placed overlay. If water seeps into the pores of the concrete, it will displace air. As the air rises to the surface to escape, it can produce pinholes and air bubbles (sometimes referred to as "fish eyes") in the fresh overlay. The second reason for using a primer is to ensure good adhesion of the overlay to the substrate. Most manufacturers recommend applying a minimum of one and in some cases two to three coats of primer, depending on the product used and the porosity of the floor. Manufacturers usually sell primers designed specifically for use with their systems.

The type of primer will determine the

Using a gauge rake to spread self-leveling overlays will ensure that the material is placed at the desired thickness. Simply preset the depth of the rake to the level needed.

method of application. Primers for some self-leveling overlays are thin enough to be sprayed onto the substrate using a pump-type sprayer.

Warning: Avoid Premature Setting

The goal when installing self-leveling overlays is to get a large volume of material down quickly. In summer or warmer conditions, you may need to extend the working time of the mixed material by using chilled or ice water to slow setting. To avoid this premature setting, keep your raw materials cool and pour the overlay during the coolest part of the day. When bagged materials containing cement are delivered or stored in warm conditions, they can retain heat for extended periods, which could accelerate the setting time once mixed.

After spray application, you should use a brush or roller to work the primer into the pores of the concrete. These primers usually require 3 to 4 hours of drying time after the final application before you can install the overlay.

Other systems may require the use of an epoxy primer—a thicker material that requires application with a notched squeegee followed by a roller to work it into the surface. When using an epoxy primer, it's usually necessary to hand broadcast sand into the wet epoxy to ensure adhesion of the overlay. Epoxy primers can take 12 to 24 hours to dry depending on jobsite conditions, such as air temperature and humidity.

Mixing

To mix overlay materials for a small to moderately sized project (2,500 square feet or less), I use several 30-gallon drums as the mixing vessels because they are large enough to accommodate two to three bags

of material. Determine the volume of water required per bag and then put all water in the clean bucket first.

Always add the exact amount of water specified by the manufacturer. Adding too much will give you a mix that's too soupy, and adding too little will make the material too stiff to self-level properly.

It's a good idea to prewet the bucket so dry material does not accumulate on the sides. Be sure to dump out any residual water before putting the required amount of water in for mixing. Once the premeasured water is in the bucket, you can add the dry overlay mix while simultaneously mixing with a power drill fitted with a mixing paddle. (Self-leveling overlays generally require 3 to 4 minutes of mixing time.)

To mix materials for larger jobs, you can still use buckets, but you'll need more of them—as well as more personnel for mixing and pouring. On one 7,000-square-foot

DECORATIVE CONCRETE INSTITUTE

On small to mid-sized projects, you can mix the components for your self-leveling overlay in 30-gallon drums and then pour the mixture onto the surface. Always add the precise amount of water specified to avoid a mixture that's too soupy or too stiff to self-level properly.

project, I had five mixing buckets going simultaneously, mixing at 1-minute intervals. Working in this fashion enabled us to pour almost continuously, expediting the job.

For mixing materials, I recommend using a top-vented industrial drill running at a minimum of 650 rpm, but avoid mixing at too high a speed, or you could end up whipping unwanted air into the mix. Always have a backup drill in case your primary drill burns out. There are numerous types of mixing paddles available. My favorite paddle for mixing self-leveling overlays is similar in shape to a football.

On large commercial projects, it's often more efficient to use a grout-type continuous mixing pump with an attached hose for placing the overlay. As a backup in case of pump failure, be sure to keep two or three mixing buckets and drills on hand. When using a pump, you usually must prime the delivery hose with either water or slurry made of cement and water. There are also prepackaged materials specifically for priming pump hoses. Make sure you have a bucket or vessel at the end of the hose to capture the slurry so it won't contaminate the floor surface.

For more information on setting up a mixing station, review Chapter 13.

Placing the overlay

Installing self-leveling overlays requires skill and a well-organized plan of attack. You have a very limited working time, ranging from less than 10 minutes to 45 minutes depending on site conditions. You don't want to end up

with a cold joint—a visible delineation that forms when a section of freshly placed overlay hardens before you have a chance to pour new material against it.

To avoid this eyesore, establish predetermined starting and stopping points. A simple method is to use adhesive-backed foam weather stripping, adhering it to the concrete floor surface as a bulkhead precisely where the overlay placement will terminate. (For better adhesion, apply a strip of fabric tape to the floor and then install the weather stripping over it.) If you prefer more decorative starting and stopping points, use stainless steel or brass divider strips or natural materials such as marble, slate, or stone. You can affix them to the floor with construction adhesive or fasteners.

The primary tool needed for spreading the overlay material onto the floor is a gauge rake, preset to the desired overlay depth. The gauge rake determines the required thickness and also spreads the material into place. I like to pour in parallel strips 3 to 4 feet wide, working from side to side.

Shortly after the material has been gauge raked into place, run a smoothing paddle (a flexible steel blade) across the surface. The smoothing paddle helps to level the overlay to a uniform depth. It also breaks the surface tension and allows any entrapped air to escape. In hard-to-access areas, such as around drains and or plumbing lines, it may be necessary to hand trowel the material level.

Color and design options

Self-leveling overlay mixes are generally offered in two colors: gray or white. But by adding pigments to the mix, your shade options are virtually unlimited. Some manufacturers offer pretinted overlay mixes, with the coloring agents already in the bag. Others can provide packages of tint that you add during the mixing process. Many companies offer dozens of standard colors to choose from and some can even custom mix colors.

Self-leveling overlays can be further enhanced by chemical or water-based stains, dyes, or tints after the surface hardens. With a little creativity, it's possible to create unique multicolored works of art that can't be duplicated with other flooring materials. For more information on the various coloring options, review Chapter 20.

Self-leveling overlays can be left seamless (except at control and isolation joints) or used as a canvas for sawcut or engraved designs. Using a variety of sawcutting tools affixed with diamond blades or special engraving tools designed for concrete, it's possible to cut almost any pattern or graphic into the overlay surface, even something as elaborate as a company logo.

You can also incorporate decorative inlays, such as strips of wood or metal, by adhering them to the base concrete and then pouring the overlay to the level of the inlay.

Spray systems are popular toppings for pool decks because they produce a decorative nonslip surface. Using a light-colored pigment will produce a reflective surface that stays cooler for walking on in bare feet.

CHAPTER 16

SPRAY-DOWN SYSTEMS

Spray-down resurfacing systems have their own distinct look and purpose. Although they are put on as thinly as most trowel-down microtoppings (no thicker than 1/8 inch), they are generally not as smooth and uniform in appearance because they are pneumatically applied. This results in interesting textural variations.

Because of their versatility, spray-down systems have become a top choice for transforming gray, unsightly concrete. Not only are these overlays durable and skid-resistant, they also offer endless decorative possibilities through the combination of various coloring, patterning, and texturing methods. Spray-down systems also work well with paper or plastic stencils, permitting the creation of intricate borders, custom designs, and logos.

Spray systems are ideal for projects where complete concrete removal and replacement is not an option, either because of cost or the risk of damaging surrounding areas. Popular applications include pool decks, patios, driveways, sidewalks, and balcony decks. They even lend themselves well to vertical applications such as retaining walls or concrete foundation walls.

Preparing the surface

As with the other overlays described in this guide, the surface must be prepared properly for spray-down systems to perform to their fullest potential. Most manufacturers of spray systems agree that some type of profiling, either chemical or mechanical, is required. If you plan to use

acid etching as a profiling method, be sure to neutralize any remaining acid residue with a solution of water and ammonia or baking soda.

Chips, spalls, and minor cracks in the existing concrete surface will also need to be repaired. For more details on making repairs, turn to Chapter 10.

Applying primer

Some spray-down systems require priming of the prepared substrate before topping installation. Check the technical data sheets for the manufacturer's recommendations for application and coverage rates. Most specifications recommend brushing the primer onto the surface.

If the system does not require a primer, you will usually need to prewet the concrete substrate to obtain a saturated surface dry (SSD) condition. This will improve the quality of the installation by aiding in the adhesion of the topping and extending the working time. In order to fully saturate a dry substrate, constant wetting for a minimum of 4 to 6 hours is necessary (and in some cases longer for highly absorbent concrete). A soaker hose or sprinkler is an effective way of wetting the surface.

Mixing

Spray-down systems are available in both single- and two-component types. Single-component systems require only the addition of water to the dry ingredients while two-component products require the addition of a liquid polymer.

Mixing procedures are similar to those used for microtoppings. First, add the proper volume of liquid (either water or polymer

additive) to a 5-gallon bucket or other mixing vessel. Then slowly add the powder component to the liquid while simultaneously mixing with a paddle drill until thoroughly blended. Use a drill with a minimum operating speed of 650 rpm, and always keep a spare drill on hand as a backup.

Special equipment needs

To apply a spray-down system, you'll need to invest in a hopper gun (sometimes called a splatter gun) and an air compressor to power an attached spray gun. Most splatter guns require a constant air feed of 5 to 7 cubic feet per minute (cfm) at a pressure of 20 to 25 psi.

You place the mixed material into the gun's hopper, where it's gravity fed to the sprayer. There are different styles and sizes of hopper guns, but most can hold 1 to 3 gallons of mixed material. Sprayer tip sizes also vary. It's important to choose one with an orifice large enough that it won't clog but not so large that it spews too much material onto the surface. Average tip sizes range from 3/16 to 1/2 inch.

Precut stencils are available in an endless array of shapes and patterns. Some manufacturers can cut the stencils to order, working from customer-supplied designs and motifs.

After all of the necessary substrate preparation has been completed, apply a base coat which will then become the color of the grout joints.

To use a precut paper stencil to form a brick-patterned border, unroll the stencil over the base coat after it dries and then spray on the texture coat in a contrasting color.

Use a random motion to spray the material straight down onto the substrate rather than at an angle. You should hold the sprayer 2 to 3 feet above the surface.

If you intend to install large volumes of spray texture, I recommend getting a self-contained hopper gun with a built-in air compressor. These units are easier to wheel around the work site since you don't need to drag around a separate air compressor.

Because you'll be spraying material, it's especially important to wear a face mask and eye protection. Refer to Chapter 22 for a list of other basic tools, equipment, and personal protective gear you'll need to apply spray-down and other resurfacing systems.

Applying the base coat

After allowing the primer to dry for the recommended time (or once the floor is saturated surface dry), the next step is to put down a base coat of the spray system. This coat will help to fill in minor pits and hairline cracks and improve the bond of the texture coat.

The other purpose of the base coat is to ensure complete coverage and better long-term durability of the overlay, especially on surfaces exposed to vehicle traffic, such as driveways. When you apply the final texture coat with the hopper gun, you may not obtain 100% coverage (depending on the spray tip used), and small areas of the underlying concrete could remain exposed unless a base coat has been applied first.

The quickest way to apply the base coat is with a rubber or metal squeegee, but you can also use a hand trowel. The goal is to spread the base layer as smoothly as possible, without leaving ridges that could show through in the texture coat.

Depending on jobsite conditions, a minimum of 2 hours drying time is usually necessary before the second coat can be applied. In cool or humid weather, longer drying of the base coat may be needed. Check the technical data sheets for recoating times.

Applying the texture coat

To apply the texture coat, use a random

Spraying Over a Stencil

When spraying the topping over a stencil, it is especially important to spray straight down over the top of the stencil. This will help to prevent any movement and keep overspray from bleeding under the stencil and blurring the pattern lines.

motion to spray the material straight down onto the substrate rather than at an angle. You should hold the sprayer about 2 to 3 feet above the surface, depending on the tip size and the amount of pressure being applied.

A popular spray-applied texture is a knock-down finish, so called because it involves "knocking down" with a trowel the high peaks that result from spray application to achieve a lightly textured surface. This is a two-person process, with one person spraying and another wielding the trowel. On surfaces where greater traction is needed, such as sloped driveways, some installers skip the knock-down process and leave the heavier texture obtained from spraying.

Color options

The most commonly used method of coloring spray systems is with an integral pigment. Most manufacturers offer a broad palette of colors to choose from. Some systems are available pretinted while others can be integrally colored during mixing.

But you don't have to stick with a monotone color scheme. Color layering can really make your work stand out from the ordinary. Some installers will accent the new surface with topically applied color, such as chemical or water-based stains and tinted sealers. Another way to layer color is to apply two coats of the spray-down material in contrasting colors. A technique I use is to apply the first coat of base color, and then wait until the surface is dry enough to walk on. Then I take a different accent color of spray material and lightly mist it over the surface. To achieve subtle accenting, I use a small tip size and a higher spray pressure. For more coloring tips, turn to Chapter 20.

Using stencils

Another way to set your work apart is to use decorative stencils to inlay or emboss designs in the overlay. The most common type of stencils used in conjunction with spray-down systems are made of plastic-coated fiberboard (laminated paper) and are offered in hundreds of patterns. Some manufacturers can produce custom stencils to replicate customer-supplied art, such as line drawings, monograms, logos, and fabric or wallpaper samples. Typically, these custom stencils are made of adhesive-backed rubber or vinyl and cut with a plotter. You can even make your own stencils by simply cutting Masonite or plywood with a jigsaw into a design. Whatever the stencil

Troubleshooting Pointers

In hot weather, spray-down materials may set too quickly, making them difficult to work with. Try to apply the overlay during the coolest part of the day, and keep your raw materials cool to help extend the working time. Also, don't try to tackle too large of an area. Section it off into segments that you can install in a controlled fashion using existing joints as starting and stopping points.

Fixing minor blemishes with spray systems is usually not a difficult task. If the material has already set, you can make repairs by mixing up another bucket of the spray material (in the same color) and then using a paint brush to stipple or flick it onto the affected area. Use a technique that imitates as closely as possible the texture achieved with the hopper gun.

material, make sure it's thick enough that it won't tear during removal.

For driveway or patio surfaces, brick- or stone-patterned stencils are a popular choice. First, I apply a colored base coat, and then once it dries, I apply the stencil to the base layer followed by spray application of the texture coat in a contrasting color. Once the stencil is removed, the underlying base coat color appears as a different colored grout joint. If you plan to apply stain accents to the overlay, it's often easier to do the staining before removing the stencil so the color underneath is not affected. Water-based stains work great for this type of application.

Usually, it's safe to remove the stencil after several hours, depending on site conditions. A simple check for determining if the stencil is ready for removal is to gently lift a corner from the surface. If the material that has accumulated on top of the stencil flakes off, then the stencil is ready to come up. If the material adheres to the stencil, leave it in place a bit longer. Removing the stencil too soon could cause the edges of the pattern to ravel. However, do not leave the stencil in place overnight. As the overlay hardens, it could lock down the stencil, making it difficult to remove.

With microtoppings, an unlimited range of colors and decorative effects are possible. This impressive microtopping installation was created by decorative scoring enhanced by rich stain accents.

CHAPTER 17

MICROTOPPINGS

Microtoppings (also called skim coats) derive their name from the term "micro," meaning very small, or in this case, very thin. But their slim physique (ranging from paper thin to roughly 1/8 inch) in no way diminishes their versatility.

Microtoppings can be used in a multitude of applications, both indoors and out. They offer an unlimited range of colors and textures, and because they are applied so thinly, they can even go on wall surfaces to achieve a look similar to Venetian plaster. Another plus: These systems often are lower in cost than self-leveling or stampable overlays because less material is used.

System types

Microtoppings fall into two general categories, distinguished primarily by how they are applied and the overall thickness. The most commonly used microtopping is put down by trowel or squeegee. This type usually requires a minimum of two coats, and in some cases three to four coats depending on the condition of the subfloor and the desired look. The first coat—called the body or structure coat—will contain coarser sand. The next application will be a finer sanded version of the same product. Some manufacturers even have superfine mixtures for use as a finish coat.

The second type of microtopping— sometimes called a semi self-leveler—is a cross between the traditional trowel- or squeegee-applied version and a self-leveling overlay. These products have self-leveling properties but are usually applied at a maximum thickness of only 1/8 inch (versus 1/4 to 1 inch thick for the typical self-leveling overlay).

Preparing the surface

Despite the differences in how these two main types of microtoppings are applied, the procedures for preparing the substrate are similar for both. Because these skim coats are so thin, the surface generally doesn't require aggressive roughening—a concrete surface profile between 2 to 4 is usually sufficient. (For more on surface profiling, review Chapter 9.)

Keep in mind that when putting down such a thin layer of material, you want to avoid making the concrete substrate too rough because the texture could mirror up through the finished topping. Some manufacturers recommend preparing the surface by shotblasting while others suggest acid etching, dustless grinding, or even light sanding using a floor buffer with sanding discs. What all manufacturers are adamant about, however, is a perfectly clean substrate free of any debris or dust— in other words, you should be able to eat your lunch off the floor!

Once the proper surface profile has been obtained per the manufacturer's recommendations, you can address any surface imperfections, such as cracks, chips, or spalling. Because microtoppings have some elasticity, I have had good success using a semi-rigid repair material, such as a polyurea, to fill joints and cracks. However, don't use a filler that's too flexible, like an elastomeric caulk, because it might inhibit adhesion of the microtopping. If there are numerous cracks in the substrate, I will often

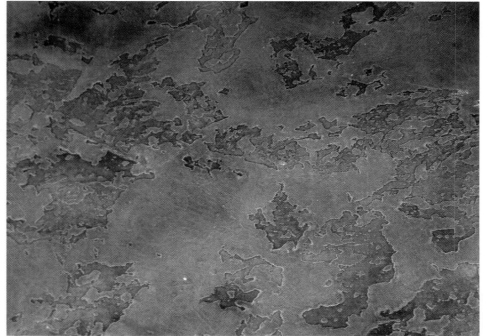

ADOBE COATINGS

Applying successive layers of microtoppings in contrasting colors results in eye-catching multi-tonal finishes. The trick is to spread the layers thinly so the underlying color from the previous layer shows through.

Here a stencil is used to create a border with the same color scheme as the multi-toned polished concrete floor it accents. A small roller is a great way to apply a microtopping over a stencil because it provides complete coverage without dislodging the stencil.

incorporate them into the pattern of the finished topping, as described in Chapter 10 on crack repair.

To fill in shallow surface imperfections, I get good results using the more coarsely sanded base coat version of the microtopping. If you have any chips and gouges deeper than 1/8 inch, fill them with a repair mortar or hydraulic cement patching compound, such as anchoring cement. These products can be applied thickly yet will dry at a rapid rate with little shrinkage.

Applying primer

Many microtopping systems require

How Much Labor?

A microtopping has a longer pot life in the bucket than other overlays but a relatively short dry time once it hits the substrate. On small to moderately sized projects (300 to 1,500 square feet) one to two people mixing is usually sufficient. With two workers, one can mix with the paddle drill while the other dumps the dry ingredients into the bucket. To apply material on small jobs, usually one person can effectively work the area. If you're covering larger floor surfaces, you may need to mix a couple of batches at a time and have several people troweling to get the material down without delay.

A broom finish is one of the simplest methods of texturing a microtopping, resulting in an attractive corduroy-like effect. The topping is usually ready for broom finishing 5 to 15 minutes after application. A fine- to medium-bristled broom works best.

application of a bonding primer or slurry coat prior to installation. Some primers are ready to use straight from the bucket while others require dilution with water.

Be sure to apply the primer according to the manufacturer's instructions and abide by the recommended drying times. Typically, primers are applied with a pump sprayer and then brushed or rolled into the surface. I know of one manufacturer that recommends applying primer with a mop and putting down a generous amount on the first application. Some primers must dry at least 8 hours, followed by a light second coat of primer applied right before microtopping installation. Others are ready for the topping after they dry to the touch or become tack-free—typically within 1 to 4 hours depending on ambient and substrate temperatures.

Bond or primer coats are important for several reasons. If you install a microtopping on unprimed concrete, the overlay system could dry instantly. That's because the porous, unsealed concrete would absorb the moisture in the topping, causing it to flash set and jeopardizing its workability and adhesion. In lieu of a primer, some manufacturers recommend applying the microtopping to concrete that's prewetted to a saturated surface dry (SSD) condition to minimize absorption of moisture from the fresh topping. When the substrate has been totally sealed off with the proper amount of primer or is SSD, the moisture in the microtopping is not absorbed into the substrate. Instead, moisture stays in the mix or bleeds to the surface of the topping, which extends the working time and facilitates troweling.

Mixing

Some microtopping systems require the addition of liquid polymer to the dry cement while others are single-component products requiring only the addition of water. When mixing two-component systems, pour the required amount of liquid polymer into the bucket and then add the integral coloring agent (if using). Mix with a paddle drill until

a uniform color is achieved and then add the dry ingredients slowly to prevent clumping. Mix for a maximum of two to three minutes. (Review Chapter 13 for pointers on setting up a mixing station.)

Although mixing techniques for microtoppings are similar to those for other overlays, the volume of material being mixed for each batch will typically be less.

Self-levelers and stamped overlays, for example, require mixing large volumes of material quickly so you can apply them expeditiously. With microtoppings, it's best to mix material on an as-needed basis since a small volume of product can cover a large area. One 5-gallon pail at a time is usually sufficient. When roughly two-thirds of the bucket has been applied to the floor, the next bucket can be mixed and ready to go.

Because microtoppings tend to dry quickly, with the rate depending on the product used and job conditions, you don't want big piles of freshly mixed material on the floor waiting to be troweled unless you have the manpower to work swiftly. Only dump from the bucket what you are able to trowel or squeegee right away.

Application techniques

Application tools and techniques for microtoppings will vary, depending on the type of topping, the size of the project, and the look you are trying to achieve.

With trowel-applied toppings, the standard application tools are a handheld steel trowel

A microtopping creates a vibrant wall mural that spills onto the floor.

with square or rounded edges or a rubber "magic" trowel (a soft rubber squeegee ranging in length from 12 to 22 inches). Whether using a steel trowel or magic trowel, my preference is to use a random trowel motion, rather than sweeping half arcs. By working the trowel in different directions, I

You can achieve a smooth, highly polished finish by giving the microtopping a second pass of the trowel. If you end up with lines in the topping left behind from your application tools, use a palm sander, orbital sander, or floor buffing machine to smooth out the imperfections once the material has dried.

avoid repeating lines. The important thing, however, is to go with a troweling style that produces the final look you're after.

On large installations, it's often more expeditious to use a rubber squeegee attached to a long handle. There are many types available specifically for installing microtoppings. Some squeegees have tapered or rounded edges while others are square. Some are made of a flexible rubber while others are more rigid. Experiment with the best tool to use for the material you are working with.

With most trowel- or squeegee-applied toppings, you'll need to put down at least two coats. Check the technical data sheets for the recommended time frame between applications. Usually a minimum drying time of 2 hours between each successive lift is required. It's unnecessary to reprime between lifts if you apply the next coat within the specified time frame.

A semi self-leveling microtopping is typically applied in one lift rather than two or more separate coats, but the finishing process is more complicated. The first step is to apply the material to depth of about 1/8 inch with a gauge roller or a gauge rake set at a depth of 1/8 inch. These tools help to spread the topping evenly while ensuring a uniform depth, but they aren't designed to move large volumes of material. When pouring the freshly mixed topping from the bucket onto the floor, apply it in a uniform ribbon rather than dumping it in one area. This will make it easier to spread properly. (It's best to have one person do the pouring while another person uses the gauge roller.) Try to spread the semi self-leveler in sections 2 to 4 feet wide, making sure to avoid cold joints (distinct lines that form between sections that dry at different times).

Once applied, these systems usually require at least one troweling, and in some cases two depending on the degree of polish you are trying to obtain. The first troweling should begin within 10 minutes after you spread the microtopping, depending on the ambient temperature as well as the temperature of the substrate and the mixed topping. (Remember, the warmer the temperature conditions, the faster the material will set.) Again, I like to use a random trowel motion to smooth and level the surface.

If you want to produce a hard-troweled surface similar to the look and feel of smooth, hard-troweled concrete, then a second troweling will be necessary. The timing for the second troweling varies depending on the size of the project and temperature conditions. I find that the most opportune time to second trowel is when 75% of the surface bleed water has evaporated, leaving behind patches of damp areas on the surface. Be sure to kneel on spiked kneeboards or wear cleated shoes when troweling to avoid leaving foot or knee prints in the fresh topping.

Coloring Options

To achieve distinctive multi-tonal finishes with microtoppings, I like to combine different coloring mediums such as integral color, stains, dyes, and tints. Another technique is to apply successive layers of microtoppings in different colors. On one project, for example, I applied a base coat of microtopping in a light gray. Two hours later, I applied a finer version of the

Because microtoppings are applied so thinly, they are ideal for use with adhesive-backed stencils to create intricate borders and other designs.

Create your own signature finish by using the trowel to produce distinctive texture variations. One interesting effect is to burnish the microtopping by running the trowel over it several hours after installation.

microtopping in a rich dark brown, and then waited another 2 hours before applying an even finer version of the mixture in a light cream color. The trick is to spread the fine and superfine applications very thinly so the underlying color shows through from the previous layer.

If you plan to use chemical or water-based stains or dyes, check with the supplier of the microtopping as well as the supplier of the coloring product for any compatibility issues and for recommendations on the best time for application. Because microtoppings are so thin, a stain or dye could soften the topping if applied before the material has cured sufficiently.

When applying stain, a simple method is to use a pump sprayer to mist it onto the surface, while simultaneously brushing with a soft- or medium-bristled push broom. Follow with one last mist of stain over the area to soften any broom marks. To avoid leaving a spray pattern, I use a sprayer with a conical tip rather than a fan tip and move the sprayer in a random fashion. For other tips on accenting overlays with color, see Chapter 20.

Decorative effects

One of the most appealing aspects of microtoppings is the versatility in texture you can achieve by using different troweling motions. How you work the trowel or squeegee determines the degree of texture. You can create your own signature finish by using the trowel to produce distinctive texture variations. If you prefer less texture, a rubber magic trowel is ideal for producing silky smooth finishes.

In most cases, the structure or body coat is applied as a base, and the second and third layers can become the texture coats if desired. Sometimes I will splatter on the second layer, either by simply letting material dribble onto the floor or flicking it on with a splash brush.

At the proper time, I will then come back and trowel the raised areas back down to level. Those of us in the trade often call this a "skip trowel" or "knock-down" finish.

Using adhesive-backed plastic stencils with microtoppings permits an unlimited range of patterns and designs, ranging from intricate decorative borders to corporate logos. Stencils also work well with color layering. On one project, I created a stunning stenciled border by applying an integrally colored base coat and then giving it a wash of green dye. I stuck the decorative adhesive-backed stencil to the first coat of dyed microtopping, and then rolled a second coat of microtopping tinted beige over the stencil. Two hours later, I applied a dark gray microtopping with a sea sponge to produce interesting color and texture variations. After allowing the second coat to dry, I removed the stencil to reveal the decorative pattern left behind in the green base coat (see photos on page 84).

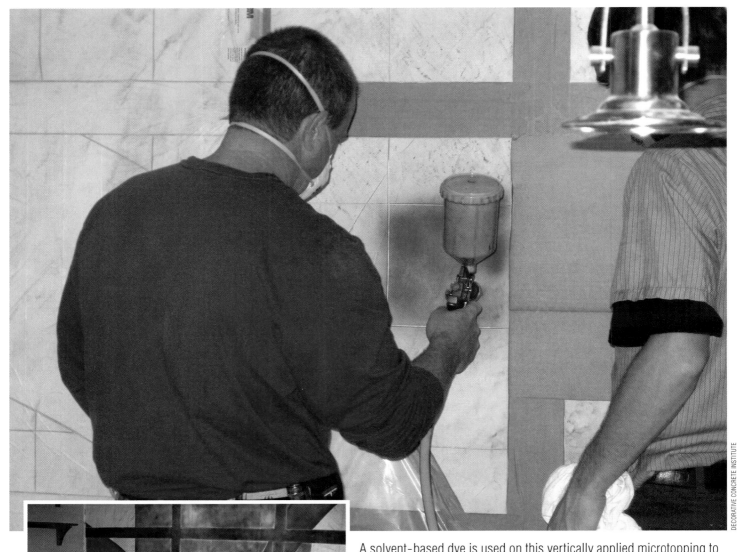

A solvent-based dye is used on this vertically applied microtopping to provide bright, bold coloring.

Warning: Dilute Stain First

If you're applying a traditional acid-based chemical stain, a little goes a long way on microtoppings, and straight stain can produce colors that are too bold. I recommend diluting the stain with water. The ratios will vary depending on the desired effect. Generally, adding two to three parts water to one part stain will work, but I have used as much as 30 parts water to one part stain to achieve the desired color level. If the microtopping is highly absorbent, I recommend lightly misting the surface with water (leaving no standing water) prior to stain application to help slow down the rate of stain absorption so the color won't be overly intense.

Microtoppings can be finished with a variety of tools including hand trowels, floats, and rubber squeegees, depending on the texture you're trying to achieve. How you work the tool also makes a difference. For example, holding the trowel flat as opposed to cocked leaves a light texture similar to smooth slate.

Troubleshooting Pointers

Although you can layer microtoppings to obtain a thicker surface (up to 1/4 inch), they are designed to be applied in ultrathin lifts. They won't perform well if you apply them too thickly or use them to fill in deep chips or gouges. If a thicker build is needed, go with a self-leveling or stampable overlay.

Drywall dust, curing compounds, asphaltic adhesives, laitance, or other contaminants on the substrate surface are the cause for many microtopping failures, as well as the failure of other resurfacing systems such as stamped or spray-applied overlays. Despite claims from some manufacturers that these types of materials will adhere to anything, you should make sure the concrete slab is spotless before applying your product.

It is also imperative to keep any source of air from moving across the surface when you're applying the topping to avoid rapid drying—especially with a semi self-leveling product. This caution applies to both indoor and outdoor projects.

On one interior flooring project, for example, I installed 1,500 square feet of microtopping that went down beautifully, except in an area where I had left two doors open on either side of the building. The cross draft caused the material to skin over in that area. Fortunately, I discovered the problem in time to avert disaster. As a last resort in a situation like this, you can lightly mist the surface with water to help provide enough cream for finishing. If you are working in a newly erected building that is not completely enclosed, you may need to erect temporary windbreaks of plywood or plastic at doorways and windows.

Even if you are as meticulous as possible with the trowel or squeegee during microtopping application, you can end up with distinct lines in the topping left behind from your application tools. If this is not the desired effect, use a palm sander or orbital sander to smooth out these imperfections once the material has dried (or the next day). On large areas, a floor buffing machine with a 100-grit sanding screen can be used.

CHAPTER 18

STAMPED OVERLAYS

Stamped overlays are an excellent option for completely rejuvenating drab interior or exterior concrete, especially on projects where removal and replacement of the concrete would be cost prohibitive. These overlays not only offer all the aesthetic attributes of traditional stamped concrete, but are also less time- and labor-intensive to install. In retrofit situations, a stamped overlay makes it possible to restore the concrete without the mess of demolition and risk of damaging surrounding structures, such as a swimming pool coping or stair and wall surfaces.

Another distinct advantage of stamped overlays is their ability to mimic other paving materials. The stamping mats or texturing skins used to imprint these overlays are available in dozens of patterns, allowing you to duplicate the beauty and texture of natural stone, brick, slate, cobblestone, and even wood planking. And in many cases, stamped overlays are less expensive than the materials they replicate.

System types

One of the most critical aspects of properly installing a stamped overlay is knowing what and what not to expect from the material you're applying. In today's marketplace, you'll find a plethora of stampable overlay systems. However, each product contains proprietary materials, which means the guidelines for mixing and installation will also differ.

How Thick?

Stamped overlays are typically applied at a thickness of 1/4 to 3/4 inch. The final thickness will be dictated by the profile of the stamping mat or texturing skin you're using.

As a general rule:
Measure where the pattern indent on the tool is deepest, and then apply the topping at a thickness twice that measurement. (Example: Deepest indent is 1/4 inch; apply overlay 1/2 inch thick).

Procedures that work great for one system may not work well for another.

Some stampable overlays are single-component products requiring only the addition of water to the dry ingredients while others are two-component systems requiring a measured dosage of liquid polymer. With some systems, you may even need to add a third component, such as cement or sand. Deciding which type of system to use will depend on a number of factors, as discussed in Chapter 1.

Preparing the surface

Regardless of the system you choose, proper preparation of the concrete substrate will be essential to obtaining the best performance. This includes taking into account any moisture-related issues (see Chapter 8), repairing existing cracks (covered in Chapter 10), and profiling the concrete. Because stamped overlays must be applied thick enough (1/4 inch or greater) to take an imprint, mechanical profiling methods such as shotblasting are usually recommended, as discussed in Chapter 9. If shotblasting is not an option, some manufacturers recommend rotary scrubbing and then acid washing. If using an acid wash, be sure to neutralize the acid and remove all residual solution by pressure washing.

Applying primer

As with all the resurfacing systems discussed in this guide, you should never just mix the materials for a stampable overlay and then apply the mixture to a raw concrete surface. With most systems, you'll need to apply a primer or slurry skim coat to act as a bonding agent and to prevent the concrete from absorbing moisture from the freshly placed overlay, causing it to flash set. Follow the recommendations provided in the technical data sheets. Many overlay manufacturers say to apply the primer and let it dry to the touch or until it's tackfree, typically after 1 to 4 hours depending on ambient and substrate temperatures. However, some manufacturers warn to never let the freshly applied primer dry before overlay installation. The number of primer coats may also vary, with some systems requiring two coats of primer, with the first coat applied the night before—or 8 to 12 hours—before overlay installation and the second coat applied just prior to installation.

In some cases, the product specifications will say to prewet the concrete substrate to obtain a saturated surface dry condition, as opposed to applying primer. When doing this, you must wet the surface thoroughly to totally saturate the concrete. Just applying a light mist of water right before overlay installation is not sufficient and could result in setting of the material before you can stamp it. In order to obtain a truly saturated substrate, constant wetting for a minimum of 4 to 8 hours is necessary. The use of a soaker hose or sprinkler is an effective way of keeping the surface wet.

Whether you are priming or saturating the surface, never leave puddles of standing water or primer. Use a paint roller, rags, or a leaf blower to dry up any damp areas before putting down the overlay.

After applying the overlay at the proper depth using a gauge rake, trowel the surface using a steel fresno, moving the blade in a slow, continuous motion to consolidate the fine aggregate. Be careful not to overwork the surface; one pass with the trowel is usually all it takes.

Mixing

Most overlay mixes are prepackaged and premeasured, but that doesn't mean the mixing process is foolproof. It's easy to make mistakes during this critical step that could undermine the success of your stamped overlay installation.

For starters, the proper volume of liquid (water or polymer) is needed to ensure the right consistency. Too much liquid in the mix could promote shrinkage cracking and a weak surface, while not enough could jeopardize the adhesion and workability of the overlay. Next, you must mix the materials thoroughly, without overdoing it. Not enough mixing could produce a stiff mix with streaks and clumps, while too much mixing could entrap air and cause the formation of unwanted bubbles in the mix.

The size of the installation will determine the best method of mixing. On small to moderately sized projects (2,000 square feet or less), I usually dump the materials in 5-gallon buckets or larger 30-gallon barrels that can hold two or three bags and then mix the components using an industrial-type paddle drill. (Make sure you have a backup drill in case one burns out during the mixing operation.)

On larger commercial projects, a mortar mixer capable of holding four to five bags (or roughly 3 to 6 cubic feet of material) will enable you to mix larger batches. To get the product from the mixer to the work area, you can dump it into wheelbarrows. For more information on setting up a mixing station, turn to Chapter 13.

Applying the overlay

A stamped overlay often requires more skill to install than some of the other toppings because of the thicker consistency of the material and the additional steps required for stamping and finishing.

The first step is to spread the material with a gauge rake—a tool similar to a garden rake but with an adjustable depth gauge for distributing topping materials at the desired thickness. It's important to perform this step in a timely fashion while the material is still viscous enough for spreading. If the mix is too dry or setting too quickly, the material will have a tendency to drag along the surface in a clump as you rake it rather than adhering to the surface. Depending on the consistency of the overlay, the gauge rake may leave behind track marks in the fresh overlay. Don't worry; you'll trowel those out later in preparation for the stamping phase.

After applying the overlay at the proper depth with the gauge rake, the next step is to trowel the surface using a steel hand trowel or fresno. Move the blade in a slow, continuous motion to consolidate the fine aggregate while bringing a layer of paste to the surface. This step is where I see many applicators make the mistake of overworking the material, trying to get it as smooth as possible. It's OK if the surface is not perfectly smooth, since you will be imprinting it later with the stamping tools. Overworking the overlay can produce "density blisters," or raised domes where air or moisture has become entrapped underneath the surface layer of mortar. One pass with the trowel is usually sufficient. If the material seems sticky and you have difficulty smoothing it out, some manufacturers suggest misting the surface with water to help lubricate the finishing trowel. Only apply a light layer of water, however, so you don't adversely affect the strength of the surface and the quality of the stamp impression.

Applying release agent

The next step is to apply a release agent to both the stamping mats and the overlay surface. Release agents come in dry or liquid form. The powdered form, which is available tinted, is sometimes called an antiquing release. Usually the stamped overlay is integrally tinted with one color and then the antiquing release is applied in a contrasting color to produce a slightly mottled, weathered appearance.

If you're using a powdered release, broadcast it over the entire overlay surface by hand or by flicking it onto the surface with a dry tampico brush about 8 inches wide. Be sure to check the product literature for the rate of coverage. Typically, one 30-pound pail is enough to cover an area of about 1,000 square feet (or an application rate of 3 pounds per 100 square feet). The goal is to apply the release in a light, uniform layer. Avoid too much buildup, which could interfere with the imprint texture.

Powdered release agents do a wonderful job of breaking the bond between the stamps and surface mortar. However, cleanup can be a messy task, especially if your project is indoors. Generally, 70% to 80% of the release agent should be removed after you stamp the overlay and allow it to cure. Removing most of the release will help you obtain subtle color tones and ensure proper adhesion of the sealer.

A liquid release agent, which is more commonly used on stamped overlays, also provides good bond breaking capabilities, and removal is usually unnecessary since most of these products will simply evaporate. The drawback is that liquid releases are only sold

You can create unique, personalized effects by experimenting with custom-made stamping tools, such as this horseshoe imprint.

clear, without any tinting. You can add your own color, however, by taking roughly 1/2 to 1 cup of a tinted powdered release and mixing it into 5 gallons of liquid release. I've learned that when doing this it's best to add the powdered release at least one day before the installation to allow the powder to dissolve completely. To apply a liquid release, use a pump-type sprayer to put down a uniform layer right before you stamp. When spraying the release, be sure to agitate the sprayer frequently by gently shaking so the pigment does not settle to the bottom of the sprayer and then when sprayed, create dark splotches on the overlay surface.

The stamping process

Achieving good results with stamping is a combination of good timing, careful planning, and sufficient tools and labor.

One of the trickiest aspects of stamping is knowing when the freshly placed overlay is at the right consistency to take the imprint. Some applicators will test for this by gently pressing the surface with a finger. If no material sticks, the time is right. Others gauge the timing by visually inspecting the surface. During the finishing operation, you will usually see some bleeding of water or polymer that settles on the surface. When this bleed water dissipates and the surface sheen becomes dull in appearance, that's usually the opportune time to start stamping.

Ideally, when you tamp the stamping mat to make the impression, there will be slight resistance. Practice on samples at your shop until you become comfortable with the best time to start.

Keep in mind that the proper timing is often a function of the depth of the pattern you are using and the size of the area to be covered. Deeper textures may require stamping earlier than light textures, especially if the placement is large. The goal is to obtain a uniform stamp impression over the entire surface before the material sets.

Well before you even place the overlay, determine which direction to run the stamping tools. If you take time to figure this out after the overlay is already down, you will get behind and the overlay will start to set. With some patterns, such as running-bond brick or cobblestone, running the pattern square is important to maintain alignment and to prevent the illusion that the pattern is shrinking or gaining in size

when viewed from different angles. Sometimes you can avoid problems with squareness by running the pattern on a diagonal. When learning how to place a new stamp pattern, it's always a good idea to practice beforehand on a layer of wet, compacted sand. If you mess up, you can simply re-level and compact the sand and start again.

Stamping in the proper sequence is also

Seamless texturing skins can be used alone or along with stamping tools to produce natural stone-like texture without distinctive pattern lines. These flexible skins are a good option for beginners because they can be placed randomly without concern about perfect alignment.

key. Generally, you should stamp in the same sequence that you placed and finished the overlay because the material that goes down first will reach the right plasticity for stamping sooner. Another good practice is to have a worker pretexture the perimeter edges of the overlay (about 6 to 12 inches inward) with a seamless texturing skin before pattern stamping. This will help to ensure that sufficient texture and full color from the release is worked into the edges. In addition, include a flexible stamp, sometimes called a "floppy," in your set of stamping tools. Highly pliable, a floppy can bend up against walls or steps, unlike a more rigid pattern stamp.

Having enough mats to cover the area you are stamping is also important so you can keep the pattern aligned and move across the surface quickly. As a general rule, have enough tools on hand to span the widest area of the overlay, plus a minimum of two additional tools to start the second row. You also need enough workers on the job to keep the process moving. A minimum of three to four people are usually required during the stamping phase.

If you start stamping at the proper time, a slight layer of mortar will squeeze up

Tips for Beginners

• If you're new to stamping, the easiest method of creating texture is through the use of seamless texturing skins, primarily because you don't have to worry about pattern alignment. I recommend that contractors start off by using the skins, and then once their skill level increases, they can move on to pattern stamping.

• As you learn the stamping process, start with smaller overlay applications so you can easily cover the entire area without concern of premature setting. In warmer conditions, consider applying only 200 to 300 square feet of material at a time. When placing the overlay in smaller sections, you can avoid a straight construction joint where the sections meet by taking a margin trowel or scraper and carving off the excess material at the edge of the stamp pattern, following the grout joint. Then, when you tie in the next placement to this joint, hand trowel the overlay gently against it and continue the stamp pattern. When done properly, you will end up with a seamless transition. As your skill level and confidence increase, you can install the overlay in larger sections.

When applying a tinted liquid release agent, use a pump-type sprayer to put it down in a uniform layer right before you stamp. Shake the sprayer frequently so the pigment doesn't settle and leave dark splotches on the overlay surface.

between the mats. This can be smoothed out with a detail roller or a rounded tool used for smoothing grout in brick joints. Have a worker devoted to detailing as the stamp mats are lifted and work progresses. If some areas are inadvertently missed, the detailer can walk back out to fix these spots using texturing skins as stepping stones (as long as the skins match the stamp pattern) assuming the material has reached sufficient set to support the weight of the worker. Or someone can do the detailing the following day by using a chisel to carefully break out the excess mortar.

As you can see, stamping can be a complicated process, but with the proper training and some hands-on experience, you'll become proficient in no time. For more in-depth coverage on stamping techniques, read *Bob Harris' Guide to Stamped Concrete*.

Removing residual release agent

After stamping, you must remove any residual powdered release to ensure adhesion of the sealer you'll apply later (as discussed in Chapter 21). Be sure to follow the manufacturer's recommendations on when it's safe to clean the release from the stamped surface. In warmer temperatures, typically 24 to 36 hours must elapse prior to release removal. In cooler weather, longer cure times may be necessary so you don't damage the surface during the cleaning phase.

Before cleaning, always test to make sure the overlay has cured sufficiently for washing.

Gently scrub or pressure wash an inconspicuous area. If no surface paste is removed, it's safe to proceed with release removal.

Some applicators use commercial detergent and push brooms to scrub away the powdered release while others prefer pressure washing. With a liquid release, removal is usually unnecessary. But you should still test for any buildup by dribbling water on the surface. If the water beads up and does not absorb into the overlay, a light cleaning should be performed.

Coloring options

With stamped overlays, an unlimited range of colors and multi-tonal effects are possible through the use of integral color, broadcast color, tinted releases, stains, dyes, and tinted sealers. This gives you the ability to precisely replicate the subtle hues of naturally weathered stone, for example, or go wild with a palette of bolder, richer colors.

Some manufacturers precolor their stampable overlay mixes while others supply packages of pigment so you can integrally color the topping during mixing. Some manufacturers say it's OK to use broadcast pigments, or dry-shake color hardeners, as another coloring option. But if you plan to use dry-shake hardeners, check with the overlay manufacturer for compatibility with their system and the timing of application. Generally, you can apply color hardener after the initial troweling of the overlay.

For more information on coloring techniques, read Chapter 20.

Troubleshooting Pointers

Since many stamped overlay installations will be on exterior surfaces, install the overlay during the coolest part of the day during the summer months. Warm and windy conditions will not only make the installation process more difficult, it can also result in premature setting and shrinkage cracking. Also, keep all of your materials cool if possible. You would be surprised how much affect the temperature of your materials has on the initial setting rate of an overlay. Read the technical data sheets for minimum and maximum temperatures for installation of the system you're using.

If the overlay gets density blisters from overworking during the troweling phase, you can usually cut the surface open with a laminated resin hand or bull float or by re-troweling with your trowel flat as opposed to cocked, allowing any entrapped air or moisture to escape.

If you're using a powdered release agent, check the polymer content of the overlay mix. With mixtures that have a high percentage of polymer, the powdered release tends to cling to the surface, making it difficult to remove the necessary amount. It may be necessary to use a dilute muriatic acid wash to remove the residual and in extreme cases, a rotary scrubber with bassine or other stiff bristles. For this reason, some overlay manufacturers may recommend using a liquid rather than a powdered release.

On large projects with a lot of area to cover, you may need to start stamping early, while the overlay material is wet, so you can complete the job before the material sets. If this is the case, the stamping tool may stick to the surface of the overlay initially, causing a suctioning effect. When this happens, never lift the tool straight up from the handles, because you could end up lifting material from the substrate and it will adhere to the bottom of the mat. First, break the bond at one corner of the mat to release the vacuum, and then lift the mat.

CHAPTER 19

VERTICAL STAMPED OVERLAYS

Cement-based overlays do not have to be confined to the floor. Advances in resurfacing products designed specifically for wall applications are making it possible for artisans to replicate the look of natural stone, block, brick, and other rustic wall finishes quickly and economically.

These lightweight blends of material are specially formulated to be applied at thicknesses of up to 3 inches without sagging. Because they go on so thickly, you can create deep rock textures and other designs using stamps or hand sculpting techniques. Accent coloring with stains or dyes completes the effect, making it possible to reproduce the multi-toned, weathered look of natural stone or aged brick.

Vertical mixes can be applied to virtually any primed wall surface, including concrete or masonry walls, cement-based backerboard, insulating concrete forms (ICFs), plaster, and even drywall. Popular applications include interior accent walls, fireplace fronts, storefronts, entryways, retaining walls, exterior privacy walls, foundation walls, and chimneys.

Preparing the wall surface

In order for a vertical mix to adhere properly, you must follow the manufacturer's recommendations for preparing the subwall. Assuming the wall is stable (without major cracks or delamination), the first step is to clean the surface to remove any contaminants. For exterior wall surfaces such as concrete and masonry, pressure washing is an effective cleaning method.

If the wall surface is extremely smooth, it may also be necessary to lightly profile the surface to improve bonding of the overlay. Methods include light sandblasting or application of a gelled acid—an etching product that will cling to the wall without dripping.

After washing the surface and allowing it to reach a saturated surface dry condition, the next step is to apply a liquid acrylic bond coat. Without priming, the wall could absorb moisture from the vertical mix, causing the material to set too quickly. After primer application, some manufacturers may also recommend applying a thin scratch coat of the vertical mix to improve adherence of subsequent coats.

Mixing

When mixing vertical overlay materials, it's imperative to get the proper ratio of liquid and dry components. If the mixture is too dry, it could crumble right off the wall because not enough moisture is present to help it congeal and cling to the surface. Conversely, if the mixture is too soupy, the material will simply run down the wall. Vertical stamp mixes from different manufacturers may have different mix ratio requirements. Check the manufacturer's technical data sheets for the proper proportions to use. You want a workable consistency with enough viscosity to hang onto the wall without sagging.

The procedures for mixing a vertical stamp mix are very similar to those for mixing the other toppings described in this guide. First, add the appropriate amount of liquid polymer to the mixing bucket or pail. Next, if integral color is to be used, add it to the liquid and then mix for about 30 seconds. Then slowly dump roughly half of the dry powder from the bag or bucket into the mixing vessel while simultaneously drilling with an industrial-type paddle drill (running at a minimum of 600 rpm). After mixing for 30 seconds, add the remainder of the material and mix until thoroughly blended (approximately 3 to 5 minutes).

On commercial projects requiring large volumes of material, you can mix the

Reinforcing Cracked Walls

If the wall surface is badly cracked or unstable, you can reinforce the wall with expanded-metal lath. I also recommend using metal lath to reinforce exterior walls subject to severe freeze/thaw conditions to ensure a stable support system for the overlay. You can attach the metal lath to the wall with an air-driven nail gun or fasteners.

ELITE CONCRETE RESURFACING

Vertical overlays can transform a fireplace front by impersonating pricier materials such as quarried stone, brick, stucco, or Venetian plaster. The lightweight cement-based mixtures will adhere to virtually any primed wall surface without sagging.

components in a mortar mixer of the desired capacity, following the same mixing sequence.

Applying the scratch coat

Some vertical mix manufacturers recommend applying a scratch coat of the material before applying the final texture coat. If this is the case, apply the scratch coat over the previously primed walls while the primer is still slightly tacky.

The scratch coat can be applied in various ways, depending on the size of the installation. One method is to use a hand trowel or a rubber sponge float (commonly used for applying tile grout) to apply the coat in a light, uniform layer no thicker than 1/8 inch. On large projects, you can cover more area faster by using a hopper gun or an industrial spray unit commonly used to apply shotcrete.

Applying the texture coat

Hand troweling is usually the preferred method of applying the texture coat, but some professionals use more creative application methods, such as taking handfuls of material and applying it to the vertical surface with a terry cloth rag—a great way to build up thicker layers of material when creating faux boulders, for example. On large projects, you can also apply the texture coat with a shotcrete-type spray unit.

When applying the texture coat, don't be concerned about smoothing it out since you'll be imprinting or sculpting it later. It is important, however, to apply the material at uniform thickness to achieve a consistent stamp impression.

If you are using stamps, the profile of the stamp pattern will dictate how thickly to apply the vertical mix. If the stamp is only lightly textured, you may get by with a 1/2 inch thickness. A stamp pattern with a deeper profile and distinct grout lines will require a thicker application.

If the final thickness you want to achieve exceeds 1 inch, I recommend applying the vertical mix in two builds. For example, to obtain an overall thickness of 2 inches, apply a 1-inch layer of the wall mix and then let it set for 15 to 30 minutes, depending on site conditions. Then go back and apply an additional inch of material. Do not allow the first coat to dry before applying the second layer. You want both applications to dry together as one coat to prevent separation.

Stamping tools and techniques

Although some stamp mats used for traditional horizontal stamped concrete are suitable for vertical stamping, you can usually obtain more realistic results with less effort by using stamping tools designed specifically for vertical applications.

These stamps are usually made from lightweight urethane foam or rubber and are typically smaller and more pliable, which allows you to press material into all crevices of the stamp and achieve deeper and more distinct patterns. You can find vertical stamps with unique designs and textures such as leaves, tree bark, animal footprints, seashells, and even fossil renderings. Stamps for deep wall patterns are sometimes made from steel and come with compatible texturing skins made of flexible rubber. First, you pretexture the wall with the skins and

then use the steel stamps to form the indents for the grout lines.

Like traditional stamp mats used on horizontal surfaces, vertical stamp sets are commonly numbered to avoid a repetitious pattern. You also must apply a release agent to the stamps to prevent them from sticking to the freshly applied wall mix. With vertical stamps, use a liquid rather than a powdered release agent. (For more information on using release agents, read Chapter 18 on stamped overlays.)

There are many variables in determining the proper time to stamp. In some cases, you will start the stamping process right after you apply the vertical mix, while other applications may require a waiting period. Primary considerations when determining the proper time to stamp are temperature, wind conditions, coverage area, and the thickness of the application. If the wall mixture starts to sag after you stamp it, then you've started the process too soon. If you see tearing at the edge of the stamp pattern, then you've waited too long.

Carving joints by hand

The "carveability" of vertical overlays can be a big advantage, especially when you want to achieve stone or masonry wall patterns with deeper reveals and grout lines than possible with stamps alone.

You can use carving tools similar to those used for carving clay, beginning your sculpting work 2 to 6 hours after placement of the final texture coat, depending on site conditions. (If material adheres to the carving tool, it's too soon and you'll need to let more time pass before carving.) If you get some raveling of the joints from carving, simply use a small paint brush dipped in water to smooth out any imperfections.

You can almost smell the coffee wafting from this hand-sculpted vertical overlay. One of the main attributes of vertical mixes is their carveability.

WORN CONCRETE CRAFTSMAN. PHOTO COURTESY OF FLEX-C-MENT

Coloring options

As with the other cement-based toppings described in this guide, there are many coloring options available to enhance your vertical stamping job. The quickest and easiest way to introduce color is to add a liquid or powdered pigment to the wall mixture during the mixing phase. Other coloring options include stains (both acid and water-based), tints, paints, and dyes.

For vertical applications, I prefer to use water-based stains or latex-based concrete paints because of their ease of application. Water-based stains and latex paints also come in an unlimited palette of colors and are easily blended to achieve custom hues.

Before adding pigments to a vertical mix, check with the manufacturer to make sure the products you plan to use are compatible with their system. If you plan to stain the wall, allow the freshly applied vertical mix to cure at least one day and in some cases longer, depending on climate conditions. Some stains, if applied too soon, could adversely affect the wall surface. It's also wise to wait a day or so before applying acrylic- or latex-based paints, to allow some of the moisture from the wall to evaporate and reduce the chance of moisture becoming trapped within the wall.

You can choose from a number of methods to apply accent color, depending on the effect you are going for. A simple technique I like to use to achieve a mottled or antiqued appearance is to apply an accent stain by spray and then immediately wipe the stain off of the high areas. You can also use traditional faux finishing techniques such as sponging or ragging, stippling with a brush, or hand painting of individual stones.

Sealing of the vertically stamped surface is recommended to help preserve the work and enhance the color. For more information on sealing, turn to Chapter 21.

Stamping tools for vertical applications are usually made from a pliable foam or rubber so they can be pressed into the overlay to achieve deep stone patterns and distinct textural effects.

To duplicate the look of weathered stone, apply an accent stain by spray, sponge, or rag, and then immediately wipe the stain off the high spots. You can also use a brush to paint individual stones by hand.

Two-Hour Rule

If you're applying a scratch coat, don't let too much time elapse before applying the texture coat.

Typically, if more than 2 hours has passed, you'll need to apply another light layer of scratch coat.

Bottom to Top or Top to Bottom?

On some jobs, you'll get better results if you start stamping from the top of the wall working your way down, while others may necessitate stamping from the bottom up. It depends on the levelness of the wall and where you want the pattern cutoff to be less noticeable. For example, on a retaining wall with a wall cap already in place, you can achieve a full course of brick, block, or stone directly under the wall cap by starting at the top and working your way down. In this case, the pattern cutoff will be less noticeable at the bottom of the wall. Conversely, if you're stamping a vertical surface that starts from a level slab, starting at the bottom is usually best.

TOP (CLOCKWISE): COLORMAKER FLOORS, RICHARD SMITH CUSTOM CONCRETE, RIVER ALLOY DESIGNS. PHOTO COURTESY OF COLORMAKER FLOORS, THE CONCRETE IMPRESSIONIST

Color can be used to establish a mood, create graphic interest, or make a floor the focal point of a room. As sources for inspiration, study the project's surroundings, the style of architecture, and the color of key elements such as walls and furnishings.

CHAPTER 20

ADDING EXCITEMENT WITH COLOR

Adding color can be the most creative aspect of installing decorative toppings, resulting in surfaces that are much more visually stimulating than bland, unadorned concrete. Depending on the project and your customer's tastes, you can choose color combinations ranging from soft and subtle to bold and vibrant. Or you can use color to replicate other materials, such as natural stone or brick.

Despite all the drama that color can offer, I encounter many installers who are intimidated by the coloring process. They say they don't have a keen eye for color, or they lack the confidence to choose colors that blend harmoniously.

With today's broad range of coloring options, you don't need to be Van Gogh to produce artistic, multihued floorscapes. This chapter offers some tips to help you unleash your creativity. Before choosing a coloring method, however, be sure to review the technical data sheets for the product you're using to determine which method best suits the application. Chapter 12 has more about the importance of reviewing the manufacturer's technical information.

Get inspiration from your surroundings

While a basic understanding of color theory is helpful, it's not essential. Some of the best color combinations come from simply observing your surroundings.

Many of my favorite color palettes have been inspired by Mother Nature, such as a breathtaking sunset or the weathered beauty of a natural rock formation. Or let the project's surrounding landscape and style of architecture be your inspiration.

For example, if you're resurfacing a concrete floor, study the colors of the walls, furnishings, and other elements in the room. Then incorporate these hues or complementary tones into your decorative overlay.

On most projects, your goal when choosing appropriate color combinations will be to achieve harmony rather than glaring contrasts. But that doesn't mean you should shy away from bold, bright colors. In some instances, color contrast can be used very effectively to establish a mood, create graphic interest, or make the floor the focal point of the room. See Sources for Color Inspiration on this page for other ideas.

Use color charts as a guide

Most manufacturers of resurfacing materials, whether skim coats, self-leveling overlays, or spray-down toppings, provide an array of standard colors to choose from—with some products available in dozens of hues. Just as you would when selecting paint for a room, use the manufacturers' color charts or chips to narrow down your options. Make sure the charts depict what the color will look like with the specific product you're using. For example, some cement-based overlays are offered in a gray or white base, with each offering a different level of intensity when colored.

Take advantage of precolored systems

Integral colors for toppings are usually achieved by mixing in dry or liquid pigments. But some systems come precolored so you don't have to add the pigment yourself. The advantage of using these products is that you can be assured of color uniformity and accuracy without having to worry about measuring out the right amount of pigment. Manufacturers that offer precolored systems usually have a broad range of standard colors to choose from. Some even offer custom color matching.

Mix in your own pigments

Some overlay systems will require you to mix in the liquid or dry pigment yourself. To help you achieve color consistency and accuracy, manufacturers often simplify the dosing process by packaging their pigments in premeasured amounts. All you need to do is add the entire contents of the package to a bag or bucket of material to obtain the desired color. Other manufacturers may provide formulas for obtaining different degrees of color intensity, such as adding 1/4 cup of pigment per bucket to produce a light tone or 3/4 cup of the same pigment for a deeper, richer hue.

Sources for Color Inspiration

- The architectural style of the buildings on the jobsite
- Nature
- Brochures and color charts from your material suppliers
- The work of other decorative concrete contractors, such as terrazzo floors or stamped concrete installations
- Trade shows displaying the latest coloring systems for decorative concrete
- Industry groups, such as the Decorative Concrete Council
- Books on decorative concrete
- Art books on color
- The photo gallery at ConcreteNetwork.com

To create greater depth of color and add contrast, hand broadcast a water-dispersible pigment onto a freshly placed overlay and then work it into the surface.

Tip: How to Remove Every Drop

If you're using a liquid pigment, here's a trick for removing any colorant that might be clinging to the container walls. Hold back a small percentage—such as a quart—of the liquid component (water or polymer) required for your mix. Once the liquid pigment has been added to the mix, pour the reserved liquid into the emptied pigment container and shake well. Then dump the entire contents into the mixing bucket. This will ensure that every drop of liquid colorant is being used.

When adding color, it's best to mix the pigment into the liquid component of the overlay product (whether water or polymer) before mixing in the powdered component. First, measure the required amount of liquid into a bucket or other vessel. Then add your pigment in the precise quantity recommended by the manufacturer, and mix with a paddle mixer. Once the color is mixed in sufficiently (usually after a minute or so), add the powdered overlay component and continue mixing. This will help to ensure a well-blended mixture free of clumps or streaks.

Straight iron-oxide pigments, both in liquid and especially in dry form, have a tendency to clump during mixing due to their large particle size. To avoid this, some manufacturers blend in a dispersing agent with a finely ground pigment that will dissolve more completely during the mixing process and achieve better color uniformity. These water-dispersable pigments can also be broadcast onto the freshly placed overlay and then troweled in to create a marbleized effect. To achieve even greater depth of color, consider using a combination of both methods, incorporating color into the base mix and then broadcasting a water-dispersable pigment onto the surface.

Experiment with topically applied color

Covering existing concrete with an overlay or skim coat lets you start with a fresh, clean canvas that is ideal for the application of topical colors, such as stains and dyes. These products aren't entirely opaque, like paint. They have some transparency, which allows you to achieve interesting antiquing or marbleized effects. Or you can apply stains or dyes to integrally colored overlays to create layers of color.

Acid-based chemical stains are most commonly used to stain cementitious overlays. But newer water-based stains are gaining in popularity as a coloring method because they are easier to work with, leave virtually no residue on the surface, and come in an unlimited range of colors. With acid-based stains, you generally have less than a dozen colors to choose from, most of which are earth tones. When using water-based stains, you should test the surface first to determine whether it's porous enough to allow the stain to penetrate, since these products will not chemically etch the surface like an acid stain. I have had good results sanding the overlay surface prior to water-based stain application

102

Premeasured units of pigment make it easy to achieve color consistency when integrally coloring an overlay.

using a rotary floor buffer and sanding screen. Always check with the overlay manufacturer to make sure this is an acceptable practice.

For greater color versatility, I like to use dyes because they allow me to achieve vibrant tones such as yellows, blues, and purples to name a few. Unlike acid stains, dyes do not react chemically with cement-based materials. Instead, they contain very fine pigments that penetrate into the concrete surface. Most dyes are packaged in concentrated form, allowing flexibility in the end color. You can use them full strength to attain greater depth of color or, depending on the type of dye, dilute them with water or solvents to produce paler shades or a light wash of color. You can also mix different colors of dye to produce your own custom hues or use dyes in conjunction with acid stains to produce a variegated look. One drawback of dyes: They aren't as colorfast when exposed to ultraviolet light as some of the other coloring systems, so they are better suited for interior floors.

Use broadcast colors and tinted release agents

As with thinner toppings, stampable overlays may be integrally colored with dry or liquid pigments as the primary coloring method. An alternative method is to apply the same shake-on color hardeners typically used for stamped concrete work. Most dry-shake hardeners are a blend of pigments, portland cement, finely graded silica sand, and wetting agents. They are broadcast onto the overlay and then floated into the surface before imprinting. In order for a dry-shake hardener to work properly, it must absorb moisture from the overlay so you can work it in.

However, some overlay systems simply do not contain enough moisture to wet out the hardener. Ask the manufacturer of the system you're installing if hardeners are a viable coloring method.

A release agent, either in powdered or liquid form, is another method of achieving color variation. These agents impart subtle color while acting as a bond breaker to prevent the stamping mats or skins from sticking to the overlay and disturbing the imprint texture. Powdered release agents are precolored while liquid releases are usually tinted on the jobsite to achieve varying degrees of color intensity.

It's also possible to add a hint of color by applying a tinted sealer to the surface. This subtle color wash is created by mixing a pigmented powder release with a solvent-based acrylic sealer. This method won't work with water-based sealers, however.

FOR MORE INFORMATION

To learn more about the use of stains, dyes, color hardeners, and release agents, read the other guides in the Bob Harris series: *Guide to Stained Concrete Interior Floors* and *Guide to Stamped Concrete*.

For details, see page 133.

Applied properly, a sealer will protect and preserve your finished artwork. Choose a product appropriate for the application and verify that it's compatible with the system you're installing.

Applying several coats of a sacrificial wax or floor finish over a sealer will further inhibit scuffs, scratches, and wear. This protective topcoat is easy to buff out and reapply as necessary.

CHAPTER 21

SEALING AND PROTECTING YOUR WORK

No decorative overlay installation is complete without the application of a protective sealer or coating.

This is the final step in the process and one of the most important. A sealer provides multiple benefits:

• It enriches the color intensity of the overlay, whether the color is integral or topically applied.

• It can add sheen to the surface ranging from satin to high gloss, depending on the product used.

• It blocks the penetration of stains from dirt, chemicals, leaves, oil and grease, and other substances.

• It improves water repellency.

• It can make the overlay easier to clean and maintain, depending on exposure conditions (see Maintaining Sealed Overlays on page 107).

Although many types of protective sealers and coatings are available, three types are most commonly used for decorative overlays and toppings: acrylics, polyurethanes, and epoxies. A general overview of these products is provided here.

Keep in mind that the type of protective sealer or coating you choose can significantly affect the life and performance of a cement-based topping (see Selecting a Sealer: Factors to Consider on page 106). Be sure to check with the overlay manufacturer for recommendations as to the appropriate sealer to use for a particular application and to verify compatibility with the overlay. Too often I have seen contractors blame surface imperfections in a topping, such as blisters or bubbles, on moisture passing up through the slab. In reality, the problems were caused by use of the wrong type of sealer or too much of it.

Choosing among the options

Most acrylic sealers are breathable, easy to apply, and inexpensive. Available in solvent- or water-based types, they are widely used on exterior toppings, such as spray-down systems. However, acrylic sealers generally are much thinner than urethanes or epoxies, so they tend to be less durable. On interior floors, it is recommended to maintain them with several coats of floor finish (wax).

Polyurethane sealers are also available in both water- and solvent-based versions and provide an abrasion-resistant surface that resists scuffs and scratches. But most of these sealers are moisture intolerant, so surfaces must be dry. Some polyurethanes can be used on both indoor and outdoor surfaces.

Epoxy coatings produce the hardest surface and bond well to cement-based materials. They provide a clear finish or are available pigmented if you wish to add monochromatic color to the topping. Because of their hardness, however, epoxies are prone to scuffing or scratching. For this reason, they are usually topcoated with a

Sealers vs. Coatings

Sealers are generally low-viscosity materials that penetrate into the pores of cement-based toppings or overlays without leaving an impenetrable film that prevents the transmission of moisture. Coatings, on the other hand, are applied more thickly than sealers and are designed to build a surface film that provides better protection and easier cleaning. However, most coatings do not permit moisture to escape from the substrate and thus shouldn't be used on surfaces that test high for moisture-vapor transmission.

Not all sealers are suitable for exterior use. Some are moisture intolerant or may yellow with UV exposure. Be sure to use a sealer, such as an acrylic, that's recommended for outdoor exposure conditions.

SUPER-KRETE INTERNATIONAL, INC.

Although sealed floors are easier to clean and less susceptible to penetration of dirt and stains, they aren't maintenance-free. Make sure your clients understand the best cleaning procedures to use to extend the life of the newly applied overlay.

CONCRETE SOLUTIONS, INC.

polyurethane or several coats of floor finish. Epoxies also have a tendency to yellow with UV exposure, so they generally are limited to interior applications. Depending on the solids content, some epoxies are nonporous and do not allow moisture in the slab to escape.

In addition to these commonly used sealing systems, some newer products are on the market that may be good alternatives for certain applications. Some manufacturers offer hybrid sealers that combine acrylic and urethane resins. These sealers, called acrylic urethanes, offer certain advantages such as easy application, quick drying times, and lower cost than traditional coatings.

Polyureas are newer high-performance coatings with excellent abrasion resistance and UV stability. A 100%-solids polyurea has some characteristics in common with polyurethanes, but the chemical makeup is different, resulting in very rapid drying times over a broad ambient temperature range. A polyurea typically dries tack-free in less than 2 hours and offers tremendous durability. However, these systems usually cost significantly more than the other protective sealers and coatings.

Sealer application methods

Acrylic sealers can be spray applied with a pump-up hand sprayer, high-volume low-pressure (HVLP) sprayer, or airless sprayer. Depending on the solids content, urethanes can be spray applied as well, but most manufacturers say it's necessary to back roll with a roller. On vertical stamped overlays, sealer is typically spray applied in light mist coats.

Although I spray apply sealer on most jobs, a roller may be a better choice on some projects, especially where overspray could be a problem. A disadvantage of rolling is that it can leave behind roller marks and pinhole bubbling if you don't use the right roller or application technique. Rolling may also cause some bleeding of topically applied dyes.

Rollers come in a variety of types and nap thicknesses. Check with the sealer manufacturer for the best roller type to use with their system. On textured toppings, a thicker 3/8-to 1/2-inch nap often works best. For smooth, nontextured surfaces, a roller with a 3/16-to 1/4-inch nap is usually sufficient.

Preferably, use a roller that's lint-free. If you can't find one, try this simple trick used by many applicators: Press a layer of tape over the entire surface of the roller, and then peel it off. This removes any loose lint that would otherwise end up on the floor. Also, read the label on the roller to verify what types of materials it can apply, such as oil- vs. water-based products, acrylics, or urethanes. In tight corners, you can use paint pads or paint brushes to cut in the sealer.

When applying epoxy coatings, which tend to be thicker than the other types, a notched squeegee rake can facilitate spreading,

DECORATIVE CONCRETE INSTITUTE

When applying a sealer in tight spots and near adjacent surfaces, use a roller rather than a sprayer to keep the material confined and to prevent overspray. Also be sure to cover the surfaces you don't want to contaminate.

depending on the mil thickness being put down. But you will still need to back roll behind the notched squeegee with a roller.

Precautions with solvent-based sealers

Some overlays and toppings are modified with polymer resins—such as acrylics, polyvinyl acetate (PVA), and ethylene vinyl acetate (EVA)—that are prone to softening if solvent-based sealers are applied in the early stages of curing. Water-based sealers, whether acrylic or polyurethane, do not have the same softening effects.

When using a solvent-based acrylic or polyurethane sealer with these toppings, be sure to give the topping enough time to cure, per the manufacturer's specifications. In addition, it's important to apply the sealer in thin coats so the solvent flashes off (or evaporates) quickly and doesn't lie on the surface for an extended time, which could soften the topping and cause blistering or bubbling. The same precautions apply to overlays topically colored with dyes. Solvent-based sealers can react with the dyes, causing them to bleed.

A good way to apply solvent-based sealer is in multiple light coats, using an airless sprayer that has the ability to apply the sealer as a mist (be sure to use a sprayer that resists solvents). You can also thin the first coat with acetone to get the sealer to flash quickly. If using acetone as a thinning agent, or any sealer that contains solvent, great precaution should be taken since solvents are extremely flammable. They can be particularly hazardous when applied by spray, since the vapor becomes airborne. Prior to sealing, be sure to extinguish any sources of open flame such as a pilot light on a hot-water heater or a lit cigarette. Solvent vapors are also harmful to breathe, so rooms must be properly vented and applicators should wear respiratory protection.

Applying a floor finish

A floor finish, or wax, is a sacrificial coating that protects the sealer from wear. Although not essential, it's often used in interior applications to protect the floor overlay from scuffs, scratches, and grime. It's very easy to buff out a coat of floor finish and then reapply more if necessary.

The most commonly used type of floor finish is a mop-down product that can be applied with a micro-fiber pad, looped-end rayon mop, or a mop kit specifically for applying floor finishes. Edges can be cut in with a sponge or rag. Never use a cotton mop to put down the floor finish because it will leave streaks. Use cotton mops

A sealer enriches the color intensity of an overlay while adding a glossy sheen.

only to clean the floor.

Most installers will apply a minimum of four to six coats of floor finish (depending on the product being used) before turning the floor over to the owner or owner's representative. The owner should be warned not to allow these original applications of floor finish to wear down to the sealer because sealers are not as receptive as a floor finish to the buffing out of scratches and scuffs.

You can purchase most floor finish products from a janitorial supply house. But first, check with the sealer manufacturer to ensure compatibility of the floor finish over specific sealers. Manufacturers of some sealers, such as chemical-resistant urethanes, do not recommend applying wax or a floor finish.

When to Apply Sealer

Check with the manufacturer's recommendations on the proper time to apply sealer. Some self-leveling overlays can be sealed within the first 4 hours if a water-based sealer or coating is used, but require a minimum of 24 hours of curing time before application of a solvent-based sealer.

Microtoppings and most other toppings usually require a curing time of at least 24 hours before application of water-based sealers or coatings versus 36 to 48 hours for solvent-based systems.

Maintaining Sealed Overlays

Maintaining properly sealed decorative overlays will depend on exposure conditions and the type and amount of traffic the surface will receive. Although a sealer will inhibit stains, the owner should still sweep and wash the surface occasionally to avoid dirt buildup. Heavily contaminated exterior surfaces can be pressure washed or scrubbed with a rotary floor scrubber and a mild detergent. For interior surfaces, wet mopping or dry dust mopping of the floor is normally the only regular maintenance needed. Application of a floor wax or polish can provide extra protection in high-traffic areas.

Over time, the sealer may begin to show some wear. If the surface begins to dull or lose its sheen, recoating with sealer can restore the luster. In high-traffic areas, such as shopping malls, it is especially important to maintain the sealed surface. Otherwise, wear patterns may begin to show, especially in colored surfaces.

A Crac-Vac with a vacuum attachment, for dust-free cutting of straight lines.

INTERNATIONAL SURFACE PREPARATION

CHAPTER 22

TOOLS, EQUIPMENT, AND SUPPLIES

To successfully place any of the resurfacing systems described in this guide, you need the appropriate tools and equipment for the task at hand. These tools of the trade will not only aid and expedite overlay application, they will also have a major impact on the ultimate finished look.

For those of you just getting started with decorative overlays, I've provided a basic list of supplies you'll need to install all the systems we've covered. If you plan to focus on a specific system or expand your business by adding another type of topping to your repertoire, I've also provided a table identifying special equipment needs by application.

Note that these lists do not include all the equipment and supplies you might need for surface preparation (discussed in Chapter 9) and crack repair (Chapter 10), since that will vary greatly depending on the condition of the existing substrate and the product you're installing. Follow the general guidelines given in these chapters, or the specific recommendations provided by the overlay manufacturer.

Without exception, the most essential supplies in your tool kit are those needed for personal protection. Be sure to review Chapter 6 for important safety precautions.

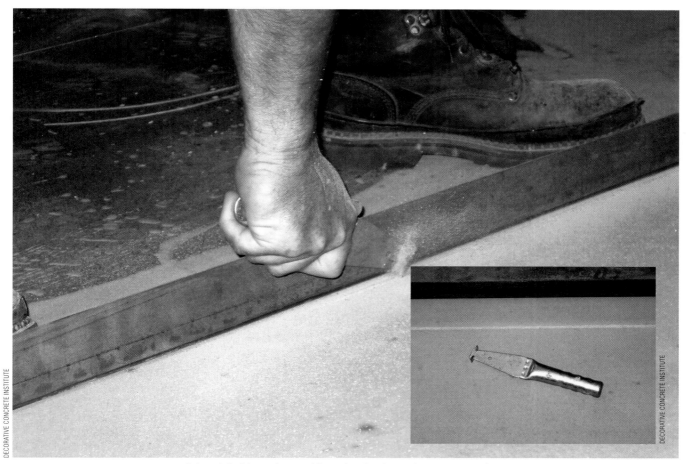

A hand-held scoring tool is a simple alternative to sawcutting.

Basic Tools & Supplies

Tape measure	To figure out the dimensions of your substrate when estimating materials. Can also be used to lay out designs.
Calculator	To calculate project square footage and material take-offs.
Pencils	Used to take off measurements and lay out designs.
Box cutters	To cut plastic or cardboard templates.
Chalkline with fluorescent chalk	For marking designs on the floor and to designate starting and stopping points.
Adhesive-backed foam weather stripping	To form starting and stopping points to prevent cold joints.
2-or 4-foot level	Used as a straight edge for layout.
Framing square or speed square	For layout and determining 90-degree angles.
Wet/dry shop vac, squeegee vac	General cleanup, water and residue removal.
Nail gun or construction adhesive	To attach metal lath or divider strips.
Soaker hose or sprinkler	To saturate dry substrates.
Pressure washer (3,000 psi)	To power wash existing substrates or remove residual release agent.
Floor scraper	Scraping up floor contaminants, removal of tile and mastic.
Extension cords	To plug in power tools and equipment or portable lighting.
Portable halogen lights	For jobs inside poorly lit buildings.
Sawcutting and engraving equipment (such as an angle grinder, walk-behind saw, Crac-Vac, Dremmel, Mongoose, and Wasp)	Cutting contraction joints, crack cleaning, and decorative scoring (see Chapter 2 on decorative finishing techniques).
Diamond tooling	Segmented and continuous-rimmed diamond blades, diamond discs for floor grinders.
Roller cages and handles, paint rollers, trays	For rolling on toppings, primers, and sealers.
Mop	General cleanup or to apply primer.
Push broom	General cleanup as well as stain and dye application.
Splash brush or tampico brush	For splatter application of materials and to broadcast powdered release agents.
Paint brushes	For detail work when applying stains or dyes.
High-volume low-pressure (HVLP) sprayer, pump-up, or airless sprayer	To apply sealers, liquid release agents, or stains and dyes.
Generator (110-220 volt, single phase)	To provide power in case of outages or when electrical outlets are unavailable.
Floor fans	For ventilation, to dry wet floors, or to speed drying times of applied materials.
Wheelbarrow or material buggy	To deliver mixed material to the worksite.
Heavy-duty mixing drills (600 rpm minimum), mixing paddles	To mix materials. Invest in at least two drills, using one as a backup.
Mixing vessels (such as 5-gallon pails or 30-gallon drums)	For mixing materials and containing water.
Measuring cups or buckets	To measure liquid components or pigment.
Water hoses	To connect to water supply, if available.
Solvents (acetone, lacquer thinner, xylene)	For thinning materials (when applicable) and for cleaning tools.
Masking supplies (tarps, rolls of paper or plastic, tape)	To protect walls and other surfaces adjacent to the work area.
Moisture-vapor test kit	To test the concrete substrate for moisture-vapor-emission levels (see Chapter 8).
Cement backerboard	To create sample boards.
Personal protective gear: Nonabsorbent rubber boots and gloves, safety goggles, hard hats, dust masks or respirators, kneepads, hearing protection	Review Chapter 6 for specific guidance.

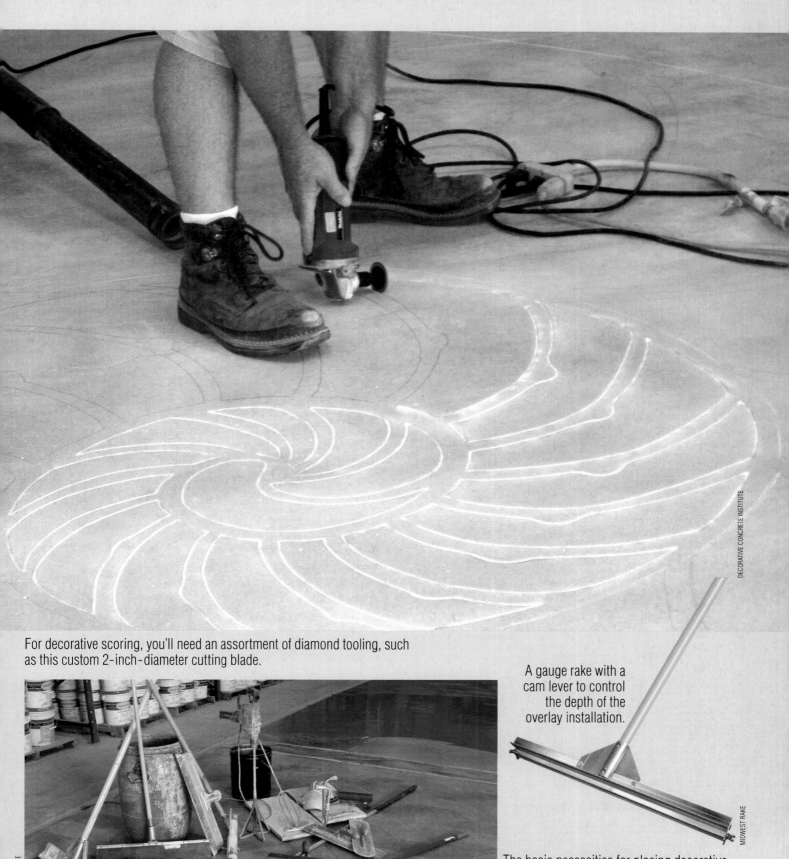

For decorative scoring, you'll need an assortment of diamond tooling, such as this custom 2-inch-diameter cutting blade.

A gauge rake with a cam lever to control the depth of the overlay installation.

The basic necessities for placing decorative toppings include a mixing drill and paddle, mixing vessels, steel-bladed tools for smoothing and troweling, kneepads and spiked kneeboards, assorted rollers and roller cages, a mop, and various handles.

Breakdown by Specialty

	Microtoppings/ Skim Coats	Spray-Down	Stamped Overlays	Self-Leveling Overlays	Underlayments	Vertical Stamped
Air compressor w/ air hose (capacity: 5-7 cfm, 20-25 psi)		X				X
Hopper gun w/sprayer tips (capacity: 1-3 gallons)		X				X
Shotcreting equipment						X
Steel hand trowels	X	X	X	X		X
Steel fresno w/extension handle						
Gauge rake			X	X	X	
Gauge roller	X					
Notched squeegee	X			X		
Stencils (plastic or paper)	X	X				
Smoothing paddle				X	X	
Porcupine roller				X	X	
Sealer*	X	X	X	X		X
Equipment & tooling for decorative scoring				X		X
Rubber magic trowel	X					
Metal or rubber squeegees	X	X				
Decorative divider strips (stainless steel, brass, wood, granite, or marble)				X		
Stamping mats/ texturing skins			X			X
Hand rollers or chisels			X			X
Carving tools						X
Tamper			X			
Integral pigment*	X	X	X	X		X
Stains or dyes*	X	X	X	X		X
Release agent*			X			X
Dry-shake color hardener*			X			
Latex-based paint*						X
Grout-type pump (capacity: 1-2 cubic yards per hour)				X	X	
Primer*	X	X	X	X	X	X
Mortar mixer (capacity: 3-6 cubic feet)**			X			X
Continuous mixer (capacity: 150-200 bags per hour)**				X	X	
Spiked shoes with non-metal cleats (such as soccer shoes), elevated kneeboards	X	X	X	X		
Reinforcing mesh or metal lath				X	X	X

*Only use a compatible product approved by the overlay manufacturer.
**Recommended for commercial projects requiring large batches of material.

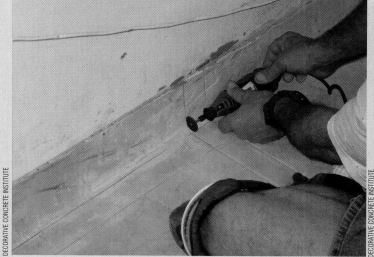

A Dremmel with a diamond tip, for detailed decorative scoring.

The Mongoose, a hand-operated sawcutting machine for making fast, accurate straight cuts as well as perfect circles.

A smoothing paddle, a flexible steel blade for smoothing self-leveling overlays and underlayments.

CHAPTER 23

TIPS FOR MARKETING DECORATIVE OVERLAYS

The main purpose of this guide is to give you the knowledge you need to install high-quality cement-based overlays. But simply mastering one or more of the techniques covered in the following chapters will not make you an overnight success. You still need to get the word out about your business.

With the growing demand for decorative resurfacing systems, some contractors believe their work will sell itself and no energy should be spent promoting their services. Not true. In fact, sometimes contractors with lesser ability but more marketing savvy have greater success attracting new customers.

Listed in this chapter are what I consider to be the fundamentals of effective marketing.

Establish a niche. If you're just starting out, I recommend perfecting one or two application methods first, such as stampable overlays or stenciling, and making them the focus of your marketing efforts. Some contractors prefer to stick with one method because they can achieve a greater level of expertise and streamline their operations. They also are more likely to gain a strong foothold in the market by establishing themselves as an expert in their chosen specialty.

Your "niche" should be a system that best suits your skill level and interests and local market demand. Once your business takes off, consider expanding the services you offer to customers.

Showcase phenomenal work. Each job should be looked at as a calling card for your business. In Chapter 3, we discussed the importance of building a portfolio that showcases your artistry. But don't overlook the opportunity to showcase your work in progress. Once you begin a job, put up a sign nearby in a prominent spot that gives your company name, phone number,

website address, and brief description of the unique services you offer. If the home or building owner doesn't object, leave the sign in place for a week or so after the job is complete. Also consider distributing informational flyers to neighboring homeowners or businesses describing what you do and where you're currently working so they can stop by to view the process— and the results. Your successful installations can be your most powerful sales tool.

Build name recognition. In addition to posting signs at your project sites, put your company logo and name on your trucks and equipment. Also spring for company shirts and hats for all your employees. This will help to build name recognition in your community and reassure customers that you and your staff are established professionals.

Open a showroom. If you have the budget and space, consider opening a showroom or design center to display samples of the various finishes and color selections you offer (Chapter 3 gives tips on preparing samples). Because overlays present endless design options, a showroom display is one of the most effective ways to introduce designers and homeowners to the wide spectrum of possibilities. Once prospects are sold on decorative resurfacing, the showroom can serve as an idea center for choosing specific colors and patterns. Your showroom doesn't have to be indoors or limited to a small array of samples. For example, think about resurfacing the entire showroom floor or even the sidewalks outside the building

with an array of decorative toppings, so customers can walk on actual installations and inspect their quality.

Exhibit at local trade shows. If you want to capture the attention of homeowners who have no idea what a decorative overlay is and how it can rejuvenate their drab concrete surfaces, consider exhibiting at a local home show where other construction trades are showcasing their services. These shows, which are often sponsored by local builders' associations or the chamber of commerce, attract homeowners who are in the mindset for renovation work and are shopping around for options. However, renting booth space and setting up an impressive display can be a major commitment in time and money. Before you take the plunge, investigate all the expenses involved and set a realistic budget.

If possible, visit the show the prior year as an attendee to scope out the exhibit hall floor plan and size up the booths of competing businesses (especially those getting a lot of traffic). If you're on a tight budget, rent a small booth for your first show and focus your efforts on networking and introducing your services. Hand out brochures and business cards, display sample boards that people can touch or walk on, and bring a portfolio of before-and-after photos. Most important, take the time to answer questions and schedule follow-up meetings with interested prospects.

Establish a professional website. Today, many buyers of products and services turn to the Internet first for information. Be sure to establish an online presence so your firm isn't

Your successful installations can be your most powerful sales tool, especially for jobs in high-traffic commercial settings.

overlooked. A website introduces new prospects to your company and helps them learn more about your background and capabilities, without the pressure of a sales call. A website can also save you time by allowing prospective customers to see a portfolio of your work before you even meet them face-to-face.

Deliver excellent customer service. Earning a reputation for reliability, as well as excellent work, will serve you well. Return phone calls, fix your mistakes, and dress professionally. Be proactive and address any concerns before they escalate into bigger problems. Being attentive to your customers' needs can set your firm apart from the competition.

Continually hone your craft. With the continuing advances in decorative resurfacing systems and techniques, there is always more to learn no matter what your skill level. Attend training seminars to master new techniques or to refine your existing skills. Network with others in the industry at events such as the World of Concrete and gatherings of the Decorative Concrete Council. Stay abreast of the latest tools, coloring agents, sealers, and other products for decorative resurfacing.

Be honest and straightforward. Tell prospects about all of the wonderful aspects of concrete toppings, but also don't neglect to tell them about the variables (as discussed in Chapter 3). Your honesty will be appreciated, and you will find that most people are fine with the realities—as long as they know what to expect.

Be patient. Becoming proficient at installing concrete overlays does not happen overnight. Likewise, building a business on a solid foundation takes time. Decide early on if you have the motivation and desire to build a business the right way—slow and steady.

It takes many years of field experience and an aggressive marketing plan to become an overnight success!

Establish yourself as an expert in a chosen specialty, such as stampable overlays, and make it the focus of your marketing efforts.

ELITE CRETE SYSTEMS, INC.

RIVER ALLOY DESIGNS. PHOTO COURTESY OF COLORMAKER FLOORS.

Overlays Can Boost Property Value

When installed properly, all the overlays described in this guide can dramatically enhance the curb appeal of a home or business, and even increase property values. I can attest to this based on firsthand experience. A spray-down overlay I installed on my own driveway helped me in the sale of my home.

The rejuvenated surface features a distinctive European fan pattern accented by faux brick borders created with stencils. Various colors were used to complement the color scheme of the home's exterior. Later, when I put my house up for sale, it sold in less than two weeks—without the help of a real estate agent. Meanwhile, other homes in the area that had been on the market for longer than eight months remained unsold.

When I asked prospective buyers what lured them to my home rather than the others, the response was unanimous: "You have a driveway that looks like a million bucks!"

Bottom line: Overlays can offer a great return on the customer's investment. Tell your clients that they, too, can get a million-dollar driveway without breaking the bank.

MODELLO DESIGNS

Each job should be looked at as a calling card for your business. Build a portfolio that showcases your finest artistic efforts.

ELITE CONCRETE RESURFACING

Once your business takes off, consider expanding the services you offer to customers, such as adding decorative wall overlays to your repertoire.

GLOSSARY

abrasion resistance – How well a concrete or overlay surface resists being worn away by friction or rubbing.

abrasive blasting – Propelling an abrasive medium (such as sand or steel shot) at high velocity against concrete to roughen, clean, or profile the surface in preparation for decorative toppings or overlays. Methods include sandblasting and shotblasting.

acetone – Common solvent. Often used as a carrier for solvent-based sealers or as a thinning agent. Considered an exempt solvent from VOC regulations.

acid etching – Application of muriatic or other acid to clean or profile a concrete surface. Used as an alternative to abrasive blasting for surface preparation.

acid stain (or chemical stain) – a stain containing inorganic salts dissolved in an acidic, water-based solution that reacts chemically with the minerals in hardened cement-based materials to produce permanent color that will not chip, peel, or flake. (Also see water-based stain.)

aggregate – A granular material such as sand, rock, crushed stone, gravel, or other particles added to cement-based materials to improve structural performance.

antiquing – A color layering technique for giving decorative overlays an aged or mottled appearance.

bleed water – Water or moisture in freshly placed concrete or cement-based overlays that rises to the surface. The presence of some bleed water can facilitate finishing.

blistering – The formation of blisters in toppings or overlays and the loss of adhesion with the underlying substrate. May be caused by moisture or moisture-vapor transmission problems or by premature application of solvent-based sealers.

body coat (or structure coat) – The base coat of a trowel-applied microtopping system, usually containing coarser sand than subsequent coats.

bond – the degree adhesion or grip of materials (such as coatings, toppings, repair mortars, or sealers) to the underlying substrate.

bond breaker – A material that prevents adhesion of materials to a concrete substrate.

bonding agent – An adhesive agent used to increase the adherence of toppings or coatings to the existing surface. (Also see primer.)

broadcast – To toss a granular material—such as dry-shake color hardener, sand, or decorative aggregate—in a uniform layer over fresh concrete, overlays, or coatings to add color or traction.

broom finish – Surface texture obtained by pushing a broom over freshly placed concrete or cementitious toppings.

build – The final wet or dry thickness of a topping or coating.

calcium chloride vapor-emission test – An ASTM test used to measure the volume of water vapor radiating from a concrete substrate over time (typically 24 hours). Often recommended before installing impermeable overlay or topping systems on projects where excess moisture is a concern.

cement (portland) – A hydraulic product that sets and hardens when it chemically interacts with water. Made by burning a mixture of limestone and clay or similar materials.

cementitious – A material containing portland cement as one of its components.

cold joint – A visible delineation that forms in concrete or cement-based overlays when placement is interrupted or delayed, allowing the material in place to harden prior to the next placement.

color layering – Applying layers of color to achieve variegated or faux finish effects, such as antiquing or marbleizing. For example, a dry-shake color hardener may serve as the base color, accented by a pigmented powdered or liquid release agent, followed by additional accenting with acid stains, dyes, or tints.

compressive strength – The maximum compressive stress concrete or cementitious overlay materials are capable of sustaining, expressed as pounds per square inch (psi).

concrete surface profile (CSP) – The degree of roughness of a concrete surface achievable with various surface preparation methods. The International Concrete Repair Institute has identified nine distinct profiles representing degrees of roughness considered to be suitable for the application of polymer-modified overlays, sealers, and coatings.

control (or contraction) joint – Sawed or tooled groove in a concrete slab used to regulate the location of cracking.

coverage rate – The area that a specified volume of material, such as a topping or sealer, will cover to a specified thickness upon drying.

crack chasing – Routing out cracks in concrete with a saw or grinder before filling with a repair material.

crack stitching – A method of repairing cracks that involves drilling holes on both sides of the crack and grouting in wire or U-shaped metal strips that span the crack.

cracks, moving – Cracks in concrete that are still moving, or active. Often they are structural in nature and continue through the entire depth of the concrete.

cracks, static – Random, non-moving hairline cracks that only affect the concrete surface. (Also see craze cracks and plastic shrinkage cracks.)

craze cracks – A series of fine, random cracks caused by shrinkage of the surface mortar.

curing – Action taken to maintain moisture and temperature conditions of freshly placed concrete or cement-based overlays during a defined period of time following placement. Helps to ensure adequate hydration of the cementitious materials and proper hardening.

degreaser – A chemical solution for removing grease, oils, and other contaminants from concrete surfaces.

delamination – A separation of a topping or coating from a substrate due to poor adhesion. Or in the case of a concrete slab, a horizontal splitting or separation of the upper surface.

density blisters – Raised domes where air or moisture has become entrapped underneath the surface layer of mortar in an overlay. Caused by overfinishing with a trowel.

dry-shake color hardener – A mixture of coloring pigments, cement, aggregates, and surface conditioning agents. Applied as a dry-shake to stamped concrete or stamped overlays to produce a colorful, wear-resistant surface.

durability – The ability of a surface to resist weathering exposure, chemical attack, and abrasion.

dyes – Translucent color solutions containing very fine pigments that penetrate into concrete or cement-based overlays. Will not chemically react with concrete like acid stains. Both water-and solvent-based dyes are available, with colors ranging from soft pastels to bolder hues such as red, blue, and orange.

engraving – The use of special tools or sawcutting equipment to score or route out patterns and designs in hardened overlays or concrete. Usually the surface is stained or colored first, so the routed areas look like grout lines.

epoxy injection – A method for sealing or repairing cracks in concrete by low-pressure injection of an epoxy adhesive.

expanded-metal lath – A sturdy but flexible diamond mesh sometimes used as a reinforcing and support system for overlays and vertical wall mixes.

featheredge – To smoothly, seamlessly blend the edge of a topping or repair material into the existing concrete.

finishing – Leveling, smoothing, compacting, and otherwise treating the surface of newly placed concrete or decorative overlays to produce the desired appearance and service properties.

filler – A flexible or semi-rigid repair material used to fill cracks or joints.

float finish – Surface texture (usually rough) obtained by finishing with a float.

fresno – A large trowel (about 2 to 4 feet in length) used for final finishing. Long handles either clip on or screw into the blade.

GLOSSARY

G

gauge rake – A tool similar to a garden rake but with an adjustable depth gauge for distributing topping materials at a preset, uniform thickness.

gauge roller – A cylindrical tool similar in appearance to a paint roller, but with rows of polypropylene spikes. An adjustable depth gauge permits application of toppings at a preset, uniform thickness.

gelled acid – A gentle etching medium thick enough to be applied by brush, permitting controlled application. Often used with adhesive stencils to lightly etch designs in overlay surfaces or to profile walls before application of vertical overlay mixes.

grinding – A mechanical surface preparation method using rotating abrasive stones or discs to remove thin coatings and mastics or slight flaws and protrusions.

H

hard-troweled finish – Surface finish obtained by using a trowel with a steel blade. Often used where a smooth, hard, flat surface is desired.

high-volume low-pressure (HVLP) sprayer – A spraying device that applies high-solids liquids at low pressure and low velocity, to reduce overspray. Use to apply sealers, liquid release agents, stains, or dyes.

hopper gun (or splatter gun) – A gravity-fed system for spray application of toppings or overlays. The material is placed in a hopper attached to a spray gun, which is powered by an air compressor. Often used to apply spray-down systems or wall mixtures.

hydration – The chemical reaction between cement and water that causes concrete or other cement-based materials to harden.

I

induction period – A waiting period sometimes recommended after mixing overlay components to permit the cement particles to fully absorb the available moisture.

integral color – A coloring agent usually premixed into fresh concrete or cementitious toppings. Can be used instead of or as a base for topically applied colors.

iron oxide – An inorganic pigment often used to color decorative toppings and coatings.

J

joint (control, expansion, or isolation) – Formed, sawed, or tooled groove in a concrete slab used to regulate the location of cracking (control joint) or to allow expansion or movement of adjoining structures. In decorative concrete, joints can also double as delineating design elements in a pattern.

K

kerf – A cut in a concrete surface made by a saw or router. (Also see sawcutting.)

knock-down finish – Achieved by applying a decorative topping with a hopper gun or by splattering onto the surface and then using a trowel to "knock-down" the material to produce a smooth or lightly textured surface.

L

laitance – A thin layer of fine, loosely bonded particles on the surface of concrete, caused by the upward movement of water. Laitance must be removed before application of a decorative topping or coating.

M

magic trowel – A soft rubber squeegee ranging in length from 12 to 22 inches. Often used to give overlays or toppings a silky smooth finish.

marbleize – To give decorative concrete or overlay surfaces the look and gloss of marble, through a combination of color layering and finishing techniques.

margin trowel – A steel trowel with a small, rectangular flat blade about 5 to 8 inches in length and a short handle. It has multiple uses, including scraping off concrete from finishing tools and applying patching materials.

microtopping – An ultrathin polymer-based decorative topping, generally less than 1/4 inch total thickness. Typically applied by trowel or squeegee, and given a textured or smooth finish. Pigments can be incorporated into the mix or broadcast onto the surface for a marbleized appearance. (Also see skim coat.)

mil – A measurement equal to 1/1,000 (0.001) inch. Commonly used to denote coating thickness.

mixing station – A designated work area outfitted with all the equipment and supplies needed to mix materials properly and efficiently.

moisture-vapor transmission – The migration of moisture vapor to the surface of a concrete slab, caused by vapor pressure differentials in the concrete and the surrounding atmosphere. Can contribute to the failure of impermeable floor toppings or coatings that do not permit moisture to escape. (Also see calcium chloride vapor-emission test.)

neutralize – To return concrete to the proper pH after acid etching, generally by washing the surface with a mixture of water and ammonia or baking soda.

notched squeegee – A rubber squeegee with notches or serrations on one or both edges. Used for smooth, consistent spreading of toppings or low-viscosity coatings.

overlay – A bonded layer of material, ranging from 1/4 to 1 inch or more in thickness, placed on existing concrete surfaces to beautify, level, or restore. (Also see polymer-modified overlay, self-leveling overlay.)

permeability – The degree to which a material will allow the passage or penetration of a liquid or gas.

pigment – A finely ground natural or synthetic particle adding color and opacity to a topping or coating.

pinholing – A surface defect characterized by pinhead-sized holes that expose the underlying substrate.

plastic – The state of a freshly mixed cementitious material before it begins to set, while it's still workable, readily moldable, and cohesive.

plastic shrinkage cracks – Irregular cracks that occur in the surface of fresh concrete or other cement-based materials soon after it is placed and while it is still plastic.

polymer-modified overlay – A cement-based overlay with polymer resins added to improve performance, wear resistance, and aesthetic qualities. Overlay manufacturers use different types of polymer resins, often blending them to produce proprietary products with unique characteristics. Many of today's decorative overlays use acrylics or vinyl blends because these resins provide excellent bond strength and UV resistance.

porcupine roller – A cylindrical tool similar in appearance to a paint roller, but with rows of polypropylene spikes. Used to help level overlay surfaces and to pop any unwanted air bubbles.

pot life – The length of time a material is useful after its original package is opened or a catalyst is added.

primer – The first coat of material applied to a concrete surface to improve bonding or adherence of subsequent coats. (Also see bonding agent.)

profile – The act of preparing a concrete surface to achieve the necessary degree of roughness (Also see concrete surface profile.)

pump-up sprayer – A hand sprayer often used to apply sealers and liquid release agents.

rebar (or reinforcing bars) – Ribbed steel bars used to provide flexural strength. Rebar come in various diameters and strength grades.

reflection cracking – The occurrence of cracks in overlays and toppings that coincide with the location of existing cracks in the substrate.

release agent – A parting agent applied to stamping tools and texturing mats before stamping to keep the mats from sticking to overlay or concrete surfaces.

rubber float – A handheld finishing tool used to give overlays a slightly roughened, sand-like finish. Ideal for surfaces requiring good slip resistance.

rubber squeegee – A finishing tool used to give overlays a relatively smooth surface. (Also see magic trowel.)

N

O

P

R

GLOSSARY

sacrificial coating – A final floor finish or wax designed to protect the sealer or topcoat from wear. Usually applied by mop or floor buffer in several coats to act as a shock absorber to scuffs, scratches, and grime.

sample (or sample board) – A small (generally 2x2-foot) representation of a decorative overlay installation, used as a selling tool or to experiment with various decorative treatments and techniques for applying materials.

sandblasting – A method of abrading or profiling a surface with a stream of sand ejected from a nozzle at high speed by compressed air. (Also see abrasive blasting.)

saturated surface dry (SSD) – Condition of concrete when the permeable voids are filled with water but no water is on the exposed surface.

sawcutting – Using a saw with abrasive blades or disks to cut joints or score patterns into hardened concrete or decorative overlays.

scarifier – Milling equipment used to clean and profile concrete surfaces or to remove existing coatings. Uses rotary impact cutters held at a right angle to the surface.

sealer – Solvent- or liquid-based material used to protect and enhance the appearance of decorative concrete or overlays.

self-leveling overlay – A flowable, polymer-modified cementitious topping with the ability to self level without troweling. Used to smooth and level existing concrete surfaces. Can also be enhanced by staining, dying, or engraving.

setting – The chemical reaction that occurs after the addition of water to a cementitious mixture, resulting in a gradual development of rigidity.

shotblasting – An abrasive blasting method using round iron shot to clean and profile concrete surfaces. (Also see abrasive blasting.)

skim coat – An overlay layer applied very thinly with a squeegee or trowel. (Also see microtopping.)

smoothing paddle – A flexible steel blade used to help level self-leveling overlays and underlayments to a uniform depth. It also breaks the surface tension and allows any entrapped air to escape.

solvent – Liquid typically used as a carrier for sealers. Highly flammable.

spalling – A breaking away of concrete at joints in floors or slabs. Typically occurs at joints that are installed improperly or don't adequately support the loads applied to them.

spiked kneeboards – Kneeboards with spikes on the bottom that elevate finishers off floor surfaces to permit easier finishing of toppings and overlays.

splatter coat – A coating or topping applied by "splattering" it onto the surface, typically by dipping a brush into the material and then flicking it. (Also see knock-down finish.)

spray-down system – A decorative overlay applied as a splatter coat or a knock-down finish to a thickness of about 1/8 inch. Often used in conjunction with paper or adhesive stencils. Available precolored or can be integrally colored during mixing.

stamped overlay – Similar to conventional stamped concrete, but can be applied to existing concrete. A cementitious topping is applied at a thickness of 1/4 to 3/4 inch and then stamped to mimic brick, slate, and natural stone. Color options include dry-shake color hardeners, colored liquid or powdered release agents, acid stains, dyes, and tinted sealers.

stamping mats – Rigid or semi-flexible polyurethane tools for imprinting stone, slate, brick, and other patterns in stamped concrete surfaces

stenciling – Using masking patterns made of paper, vinyl, or plastic to create stenciled concrete effects. Some stencils are adhesive-backed to keep the patterns firmly in place on the concrete surface while the decorative treatment of choice is applied, such as acid stains, dyes, spray-down systems, and etching gels.

surface preparation – Preparing concrete surfaces prior to resurfacing or application of a decorative coating to remove contaminants and minor defects or to obtain the necessary degree of roughness for adequate bonding. (Also see abrasive blasting, acid etching, and grinding.)

tack – The stickiness or adhesiveness of a material.

tamper – A handheld impact tool used to firmly press texturing mats into concrete or overlays while still plastic.

technical data sheet – Contains important specifications and manufacturer guidelines for product usage. Includes such data as coverage rates, recommended applications, product limitations, surface preparation guidelines, mix ratios and required mixing times, pot life, application procedures, cure times, performance data, and precautions.

texturing – Giving concrete or overlay surfaces a texture without leaving deep pattern lines.

texturing skins – Flexible skins for adding seamless textures to concrete or overlay surfaces. Generally thinner and more pliable than mats.

tint – A diluted color wash used to add hints of color to decorative overlays.

trowel – A flat, broad-bladed steel hand tool used to compact the paste layer at the surface and provide a smooth, flat finish. Also useful for applying topping or repair materials. Available in different shapes (with rounded or square edges) and lengths (ranging from 8 to 24 inches). Smaller trowels are useful for borders, work in restricted areas, or to work in accents of dry-shake color hardener. (Also see margin trowel, fresno).

underlayment – A cement-based resurfacing system used to restore structurally sound yet worn concrete in preparation for floor coverings. May also be applied to wood subfloors.

vertical stamped concrete – A decorative finish for walls and other vertical surfaces using a lightweight cementitious overlay formulated to be applied at thicknesses of up to 3 inches without sagging. While the overlay is still plastic, it can be stamped or hand carved to produce deep-relief stone or masonry wall textures. After the material dries, acid stains or dyes can be sprayed or sponged onto the surface to give it the multi-toned look of natural stone.

viscosity – A measure of the fluidity of a liquid material. The more viscosity a material has, the less it flows.

volatile organic compounds (VOCs) – Organic compounds that readily vaporize at normal room temperatures. Concrete coatings, sealers, or cleaning materials that are solvent-based generally have a higher VOC content than water-based materials. Some VOCs can be hazardous when inhaled.

wall veil – A microtopping applied to a vertical wall surface

water plug – Hydraulic cement used to fill cracks and to prevent the migration of moisture.

water-based stain – An acrylic-urethane based stain available in a broader palette of colors than acid stains. Very low in volatile organic compounds, with workability characteristics similar to latex paint. Can be applied to concrete surfaces by brush, roller, sponge, cloth, or commercial sprayer.

water-cement ratio – The ratio of the amount of water to the amount of cement in a concrete mixture.

workability – The ease of which freshly mixed cement-based materials can be mixed, placed, and finished.

working time – The amount of time available for placing and finishing a cement-based material before it begins to set. Often depends on the ambient temperature and substrate temperature.

xylene – A common solvent. Used as a carrier for solvent-based sealers and as a thinning agent. High in odor and flammability.

RESOURCES

Ready to get started with decorative concrete overlays and toppings? Here are some good resources for tools, supplies, equipment, and training. Many of these companies will ship their products worldwide.

AMERICRETE INC.

41769 Enterprise Circle North, Ste. 104
Temecula, CA 92590
Phone: 800-775-8880
Fax: 951-296-1033
www.americrete.com

A supplier of water-based, environmentally friendly resurfacing products as well as water-based stains, epoxies, urethanes, and chemical- and mar-resistant sealers.

ARCHITECTURAL ENHANCEMENTS

12906 Ventura Court #5B
Shakopee, MN 55379
Phone: 952-233-2726
www.decorativeoverlayments.com

Offers a broad range of decorative resurfacing systems and companion products including vertical and horizontal stamp mixes, ultra-smooth architectural finishes, acid stains, crack fill system, adhesive stencils, stamps, and solvent- and water-based sealers. Also conducts hands-on training seminars.

ARCUSSTONE PRODUCTS, LLC

5601 San Leandro St.
Oakland, CA 94621
Phone: 510-535-9300
Fax: 510-535-9400
www.arcusstone.com

Manufactures a proprietary resurfacing system that can be applied to interior or exterior wall surfaces to emulate cut limestone, polished travertine, and stucco. Available in a variety of standard and custom colors.

ARDEX ENGINEERED CEMENTS

400 Ardex Park Dr.
Aliquippa, PA 15001
Phone: 724-203-5000
Fax: 724-203-5001
www.ardex.com

A source for underlayments, self-leveling overlays, decorative concrete toppings, and moisture control systems.

ARTCRETE

5812 Hwy. 494
Natchitoches, LA 71457
Phone: 318-379-2000
Fax: 318-379-1000
www.artcrete.com

Supplies decorative concrete stencils in a wide range of patterns as well as texturing tools, color hardener, liquid and powdered release agents, and chemical stains.

AXSON NORTH AMERICA, INC.

1611 Hults Dr.
Eaton Rapids, MI 48827
Phone: 517-663-8191
Fax: 517-663-0523
www.axson-na.com

Carries penetrating epoxy sealers and products for repairing cracks and filling joints.

BOMANITE CORPORATION

232 S. Schnoor Ave.
Madera, CA 93637
Phone: 559-673-2411
Fax: 559-673-8246
www.bomanite.com

Bomanite partners exclusively with a worldwide network of experienced and specially trained concrete professionals dedicated to the highest standards of quality and service. These specialists and Bomanite—the company that invented the imprinted concrete process more than half-century ago—form an unbeatable, unique team.

BRICKFORM RAFCO PRODUCTS

11061 Jersey Blvd.
Rancho Cucamonga, CA 91730
Phone: 800-483-9628
Fax: 909-484-3318
www.brickform.com

Carries a full line of decorative concrete products including texture mats, sandblast stencils, acid stains, integral color, dry-shake color hardener, overlays, and sealers.

CADILLAC CONCRETE PRODUCTS

7513 Edmonds St.
Burnaby, BC V3N1B5 Canada
Phone: 604-830-1812
Fax: 604-526-1813
www.cadillacconcrete.com

Specializes in custom stamping mats and seamless texturing skins in stone, slate, tile, brick, and wood plank patterns.

CMP SPECIALTY PRODUCTS

P.O. Box 3614
Maple Glen, PA 19002
Phone: 215-672-6364
Fax: 215-672-6384
www.cmpspecialtyproducts.com

Manufactures a complete line of self-leveling and trowelable underlayments, industrial wear surfaces, self-leveling decorative toppings, and floor patching compounds.

COLOR-CROWN CORPORATION

928 Sligh Ave.
Seffner, FL 33584
Phone: 800-282-1599
Fax: 813-655-8830
www.stardek.com

Color-Crown's Stardek is a line of decorative overlay products for concrete pool decks, driveways, walkways, patios, and garages. The systems are applied by trowel, sprayer, or roller; patterns are created with the use of stencils.

COLORMAKER FLOORS LTD.

14273 Knox Way, Unit 143
Richmond, B.C. V6V2Z7
Phone: 604-244-3122
Fax: 604-244-3115
www.colormakerfloors.com

Offers a line of polished concrete toppings, cementitious overlays, concrete resurfacers, acid stains, dyes, integral coloring, and sealers exclusively for decorative concrete flooring applications.

CONCRETE COATINGS, INC.

1105 N 1600 W
Layton, UT 84041
Phone: 801-544-8771
Fax: 801-544-5896
www.concretecoatingsinc.com

Supplies acrylic and epoxy floor finishes for restoring existing concrete as well as structural repair products and acid stains.

CONCRETENETWORK.COM, INC.

11375 Oak Hill Lane
Yucaipa, CA 92399
Phone: 866-380-7754
Fax: 909-389-7744
www.concretenetwork.com

The Concrete Network provides a window to the world of concrete products, concrete services, and concrete service providers. Visitors to the site can find information on many popular concrete topics, including decorative concrete floors, concrete countertops, decorative concrete pool decks, patios, driveways, and much more. The Concrete Network's ConcreteProductsWeb.com helps contractors find suppliers of products for decorative concrete by category and location.

CONCRETE SOLUTIONS, INC.

3904 Riley St.
San Diego, CA 92110
Phone: 800-232-8311
Fax: 619-297-3333
www.concretesolutions.com

Supplies products for the repair, restoration, and beautification of existing surfaces, such as a stampable overlay system, a spray-applied polymer-modified cement for recoloring or restoring concrete, and a decorative color-flake system for producing granite or terrazzo looks.

CONCRETE TECHNOLOGY INCORPORATED (CTI)

8770 133rd Ave.
North Largo, FL 33773
Phone: 800-447-6573
Fax: 727-536-8273
www.cti-corp.com

CTI's modified-acrylic cement system for restoration of residential and commercial concrete surfaces achieves the look, texture, and color of any type of stone. Provided through a nationwide network of installers.

CROSSFIELD PRODUCTS

3000 E. Harcourt St.
Rancho Dominguez, CA 90221
Phone: 310-886-9100
Fax: 310-886-9119
www.miracote.com

Crossfield's Miracote polymer-based decorative finishes for interior or exterior concrete can be used to achieve many different looks, from intricate, artistic designs to textured antique finishes. Also included in the product line are waterproofing systems, pigmentation products, coatings, and sealers.

DECORATIVE CONCRETE INSTITUTE

8729 South Flat Rock Rd.
Douglasville, GA 30134
Phone: 877-324-8080
Fax: 770-489-4948
www.decorativeconcreteinstitute.com

Bob Harris' Decorative Concrete Institute provides consulting, education, installation, and on-the-job training to architects, artists, concrete finishers, faux finishers, general contractors, and interior designers across the U.S. and internationally. Some of the topics covered in the curriculum include decorative stamping, staining techniques, faux finishes, stenciling, design layout, decorative score cutting, and sandblasted and engraved graphics.

DECOSUP

Florida:
8232 NW 56 St.
Miami, FL 33166
Phone: 305-468-9998
www.decosup.com

Dallas:
3806 Melcer Dr.
Rowlett, TX 75088
Phone: 214-607-4084
www.decosup-dallas.com

Ships decorative concrete products anywhere in the world. Choose from acid stains, dyes, decorative overlays, sealers, and more.

ELITE CRETE SYSTEMS, INC.

P.O. Box 96
Valparaiso, IN 46384
Phone: 888-323-4445
Fax: 219-945-1982
www.elitecrete.com

Supplier of cementitious overlay systems, seamless texturing skins, stencils, antiquing stains, liquid colorants, release agents, and sealers and protective coatings.

ENGRAVE-A-CRETE

Manasota Industrial Park
4693 19th St. Court East
Bradenton, FL 34203
Phone: 941-744-2400
Fax: 941-744-2600
www.engraveacrete.com

Provides tools to score, saw, engrave, or cut decorative designs in concrete, including the Mongoose, Wasp, and KaleidoCrete system. Many standard and custom template patterns also available.

FLEX-C-MENT

1810-1 East Poinsett St.
Greer, SC 29651
Phone: 864-877-3111
www.flex-c-ment.com

Specializes in cementitious overlay materials for producing stamped concrete floors, walls, and countertops. Also sells reusable rubber stamps for producing deep-relief stone or masonry wall textures.

FOSSILCRETE
544 SE 29th St.
Oklahoma City, OK 73129
Phone: 405-601-8009
Fax: 405-602-0560
www.fossilcrete.com

Offers a unique assortment of concrete stamping tools replicating the art forms of nature. Categories include animals and tracks, fossils, plants and trees, and marine life. Also supplies a vertical stamping mix.

FRITZ-PAK CORPORATION
11220 Grader St.
Dallas, TX 75238
Phone: 888-746-4116
Fax: 214-349-3182
www.fritzpak.com

Sells premeasured, powdered admixtures packaged in ready-to-use water-soluble bags. Product line includes set retarders, superplasticizers, water reducers, accelerators, and air entrainers.

GARON PRODUCTS, INC.
P.O. Box 1924
Wall, NJ 07719
Phone: 732-223-2500
Fax: 732-223-2002
www.garonproducts.com

A source for decorative epoxy floor resurfacers as well as concrete floor cleaners, surface preparation and repair products, and crack and joint fillers.

GENERAL POLYMERS
145 Caldwell Dr.
Cincinnati, OH 45216
Phone: 513-761-0011
Fax: 513-761-4496
www.generalpolymers.com

Product line includes decorative seamless flooring, high-performance industrial flooring, waterproofing systems, and concrete repair products.

HACKER INDUSTRIES, INC.
610 Newport Center Dr., Ste. 250
Newport Beach, CA 92660
Phone: 949-729-3101
Fax: 949-729-3108
www.hackerindustries.com

Supplies cementitious underlayments for concrete and wood floors, including high-strength, sound insulating, and fire-resistant systems.

HOVERTROWEL, INC.
5048 Spruce Lane
Mohnton, PA 19540
Phone: 610-856-1961
Fax: 610-856-1920
www.hovertrowel.com

Specializes in installation equipment for floor overlays, including a patented pneumatic-driven power trowel designed for finishing polymer floor toppings and polymer-modified cementitious systems.

HTC AMERICA
P.O. Box 5077
Knoxville, TN 37928
Phone: 865-689-2311
Fax: 865-689-3991
www.htc-america.com

Sells machines, tools, and accessories for grinding and polishing concrete floors.

INCRETE SYSTEMS, INC.
1611 Gunn Highway
Odessa, FL 33556
Phone: 813-594-0197
Fax: 813-886-0188
www.increte.com

Offers stampable and spray-applied decorative overlay systems as well as integral colors and chemical stains.

INTERNATIONAL CONCRETE REPAIR INSTITUTE (ICRI)
3166 S. River Rd., Ste. 132
Des Plaines, IL 60018
Phone: 847-827-0830
Fax: 847-827-0832
www.icri.org

An industry organization devoted to providing publications, bulletins, and other guidance on concrete restoration, repair, and protection.

INTERNATIONAL SURFACE PREPARATION

6330 West Loop South, Ste. 900
Bellaire, TX 77401
Phone: 713-830-9300
Fax: 713-644-1785
www.surfacepreparation.com
www.blastrac.com
www.sawtec.com

This worldwide distributor of various brands of concrete cutting, grinding, and surface preparation equipment is a source for the Crac-Vac for decorative straight cutting, handheld and walk-behind grinders for profiling and mastic removal, and shotblasting equipment for surface preparation and profiling.

KEMIKO CONCRETE PRODUCTS

P.O. Box 1109
Leonard, TX 75452
Phone: 903-587-3708
Fax: 903-587-9038
www.kemiko.com

Kemiko's Stone Tone line of acid stains, wax, and sealer can be used to give concrete floors the look of marble or glazed stone. Also available: an acrylic-urethane polymer stain offered in a full palette of bold colors.

KEY RESIN COMPANY

4061 Clough Woods Dr.
Batavia, OH 45103
Phone: 888-943-4532
www.keyresin.com

Specializes in epoxy resin systems for concrete. Products for decorative flooring applications include clear resin finishes combined with colored aggregates, thin-set terrazzo systems, and clear and colored protective coatings.

KOVER KRETE SYSTEMS

22 N. Dollins Ave.
Orlando, FL 32805
Phone: 407-246-7797
Fax: 407-481-2261
www.koverkrete.com

Sells an acrylic cementitious overlay system for commercial and residential pool decks, patios, walkways, entries, driveways, and parking lots. Product can be textured, stenciled, or stamped. Also available: solvent- and water-based paints and sealers, acid stains, cleaners, and mastic removers.

L. M. SCOFIELD COMPANY

6533 Bandini Blvd.
Los Angeles, CA 90040
Phone: 800-800-9900
Fax: 323-720-3030
www.scofield.com

Provides engineered systems for coloring, texturing, and improving the performance of architectural concrete, including coloring admixtures, floor hardeners, colored cementitious toppings, stains, curing agents, sealers, coatings, repair products, and texturing tools. The Scofield Institute offers comprehensive technical information and contractor training.

L&M CONSTRUCTION CHEMICALS, INC.

14851 Calhoun Rd.
Omaha, NE 68152
Phone: 402-453-6600
Fax: 402-453-0244
www.lmcc.com

Produces a flowable topping for high-wear, heavy-duty floor applications as well as a broad array of products for concrete sealing, surface preparation, crack repair, and joint protection.

LIFE DECK SPECIALTY COATINGS

770 Gateway Center Dr.
San Diego, CA 92102
Phone: 800-541-3310
Fax: 619-262-8606
www.lifedeck.com

Offers an array of overlay and coating products for producing an infinite number of finishes including stone and brick patterns, knock-down textures, and smooth or broomed surfaces.

MAPEI CORPORATION

1144 E. Newport Center Dr.
Deerfield Beach, FL 33442
Phone: 800-426-2734
www.mapei.com

Produces a line of advanced concrete restoration systems including underlayments, toppings, primers, epoxies, repair mortars, additives, waterproofing materials, grouts, and accessories.

MARBELITE INTERNATIONAL CORPORATION

1500 Global Court
Sarasota, FL. 34240
Phone: 941-378-0860
Fax: 941-378-9832
www.marbelite.com

Product line includes spray cement textures, bonding resins, epoxies, retarders, color tints, polymer additives, and pigmented and clear sealers.

MARSHALLTOWN COMPANY

104 South 8th Ave.
Marshalltown, IA 50158
Phone: 641-753-5999
Fax: 641-753-6341
www.marshalltown.com

Manufacturer of finishing tools for concrete including bull floats, trowels, hand floats, edgers and groovers, kneeboards, chisels, tampers, and more.

MIDWEST RAKE

P.O. Box 1674
Warsaw, IN 46581
Phone: 574-267-7875
Fax: 574-267-8508
www.midwestrake.com

A resource for tools and other accessories for mixing and applying decorative overlays. Among the products offered: gauge rakes, smoothing blades, squeegees, mixing barrels, magic trowels, roller covers and frames, hand floats, spiked shoes, and tool handles.

MODELLO DESIGNS

2504 Transportation Ave., Ste. H
National City, CA 91950
Phone: 619-477-5607
Fax: 619-477-0373
www.modelloconcrete.com

Modellos are one-time-use adhesive masking patterns for decorative stenciling of concrete overlays. They are offered in a wide assortment of stock patterns including medallions, motifs, corners, borders, and tiles. Custom patterns and designs are also possible. Other products include water-based stains, a thickening agent for liquid coloring mediums, and a premixed etching gel.

MORTEX MANUFACTURING COMPANY, INC.

1818 W. Price St.
Tucson, AZ 85705
Phone: 520-887-2631
Fax: 520-293-8884
www.mortex.com

Carries stampable and spray-applied toppings for pool decks, walkways, patios, and other concrete surfaces. Application tools and products for concrete cleaning, repair, and renovation are also available.

PERMACRETE

1101 Menzler Rd.
Nashville, TN 37210
Phone: 615-331-9200
Fax: 615-834-1622
www.permacrete.com

Supplies, through a network of dealers, everything needed for resurfacing vertical and horizontal concrete surfaces. Overlay mixes, colorants, sealers, cleaners and degreasers, crack repair kits, adhesive templates, and installation equipment are among the products available.

PROFESSIONAL DECORATIVE CONCRETE SERVICES, LLC / RENEW-CRETE SYSTEMS, INC.

2020 Murrell Road
Rockledge, FL 32955
Phone: 407-677-6267
Fax: 407-677-7978
www.renewcrete.com

A supplier of Renew-Crete cement-based restoration systems for new or existing concrete surfaces, including a textured spray coat system and decorative stamped overlay.

QC CONSTRUCTION PRODUCTS

232 South Schnoor Ave.
Madera, CA 93637
Phone: 559-673-2467
www.qcconprod.com

Has a wide range of colorants for achieving hues ranging from subtle to bright, such as integral coloring, patina stain, dry-shake color hardener, and a penetrating water-based tinting compound. Also sells protective sealers and stripping products for removing worn sealer and finishes.

RUDD COMPANY, INC.

1141 NW 50th St.
Seattle, WA 98107
Phone: 206-789-1000
Fax: 206-789-1001
www.ruddcompany.com

Rudd's SkimStone is a proprietary blend of acrylic resins, portland cement, and other ingredients that can be applied in paper-thin layers over concrete surfaces to create a decorative textured finish. Material can be custom tinted with liquid coloring agents.

SMITH PAINT PRODUCTS

1817 North Cameron St.
Harrisburg, PA 17103
Phone: 717-233-8781
Fax: 717-232-5199
www.smithpaints.com

Smith's Color Floor and Color Wall are water-based extra-strength stains for decorative concrete floors, overlays, walls, and other porous mediums. Color Floor Gel is a viscose water-based stain for extra control when hand painting intricate patterns and crisp edges or lines.

SPECIALTY CONCRETE PRODUCTS, INC.

P.O. Box 2922
West Columbia, SC 29171
Phone: 803-955-0707
Fax: 803-955-0011
www.scpusa.com

Manufacturer of resurfacing materials for new concrete construction or renovation, including stampable, stenciled, and self-leveling systems. Other products include acid stains, sealers, and integral colors. Also offers hands-on decorative concrete training classes.

SPRAY-CRETE SYSTEMS

4520 West Linebaugh Ave.
Tampa, FL 33624
Phone: 813-249-7505
Fax: 813-249-7907
www.spray-crete.com

Proprietary resurfacing system, for use over existing or new concrete, gives surfaces a safe, slip-resistant finish and is available in light cool-to-the-touch colors. Applications include pool decks, patios, walkways, and entryways.

THE STAMP STORE

121 NE 40th St.
Oklahoma City, OK 73105
Phone: 888-848-0059
Fax: 405-525-3367
www.thestampstore.com

Offers an extensive line of texture mats
and skins in hundreds of stone, tile, slate,
brick, and wood patterns. Also supplies
stampable overlay mixes, integral colors,
chemical stains, color hardeners, release
agents, and stencils.

SUNDECK PRODUCTS, INC.

805 Avenue H East, Ste. 508
Arlington, TX 76011
Phone: 888-390-0305
Fax: 817-649-7292
www.sundek.com

Sundek's acrylic decorative concrete
coating is applied in multiple layers to
old or new concrete to achieve a textured
finish or various tile or brick patterns.
Available in 17 standard colors and
custom hues.

SUPER-KRETE INTERNATIONAL, INC.

1290 N. Johnson Ave., Ste. 101
El Cajon, CA 92020
Phone: 619-401-8282
Fax: 619-401-8288
www.super-krete.com

Offers a complete system of products
for repair, restoration, renovation,
resurfacing, protection, and decoration
of concrete substrates, including
underlayment and stampable overlay
mixes requiring only the addition of
water.

SUPERSTONE, INC.

1251 Burlington St.
Opa-Locka, FL 33054
Phone: 305-681-3561
Fax: 305-681-5106
www.superstone.com

Supplies a broad range of products for
concrete resurfacing and beautification,
including texturing mats, penetrating
chemical stains, liquid colorant, integral
color, solvent-based acrylic sealers,
and resurfacing and crack repair
polymers.

SURECRETE DESIGN PRODUCTS

15246 Citrus Country Dr.
Dade City, FL 33523
Phone: 352-567-7973
Fax: 352-521-0973
www.surecretedesign.com

Offers a spray-applied texture system
and a stampable cementitious topping
for interior and exterior surfaces.
Companion products include stencils,
stamping tools, and chemical stains.

SURFACE GEL TEK

663 W. 2nd Ave. #15
Mesa, AZ 85210
Phone: 480-970-4580
Fax: 480-421-6322
www.surfacegeltek.com

Sells a gelled acid product that works
with vinyl stencil materials for precision
decorative stenciling of concrete. It can
also be used to open up steel-troweled
concrete surfaces and to produce a
uniform etched surface for improved
penetration and adhesion of subsequent
products.

TRIANGLE COATINGS, INC.

1930 Fairway Dr.
San Leandro, CA 94577
Phone: 510-895-8000
Fax: 510-895-8800
www.tricoat.com

Triangle's White Mountain line of
maintenance and restoration solutions
for concrete includes a nontoxic
industrial-strength cleaner, stain-repellant
sealer, high-gloss water-based floor
polish, and one-step concrete stain
and sealer.

RESOURCES *continued*

USG CORPORATION
125 S. Franklin St.
Chicago, IL 60606
Phone: 800-874-4968
www.usg.com

Offers high-strength polymer-modified cementitious underlayments for both concrete and wood substrates.

VAPRECISION, INC.
2941 West MacArthur Blvd., Ste. 135
Santa Ana, CA 92704
Phone: 714-754-6141
Fax: 714-549-8245
www.vaportest.com

Supplies calcium chloride kits, tools, and technical support for moisture-vapor testing of interior concrete slabs.

VERSATILE BUILDING PRODUCTS
20420 South Susana Road
Carson, CA 90810
Phone: 800-535-3325
www.deckcoatings.com

A manufacturer of complete systems for concrete resurfacing, such as deck coatings, garage floor coatings, industrial floor coatings, and decorative cementitious overlays. Also supplies concrete stains, sealers, and repair products.

SPECIAL THANKS TO THE FOLLOWING CONTRACTORS FOR PROVIDING PHOTOGRAPHY:

ADOBE COATINGS, INC.
55 N. Sunaire Dr.
Mesa, AZ 85205
Phone: 480-830-0070
Fax: 480-830-8515
www.adobecoatings.com

BAY AREA CONCRETES, INC.
4179 Business Center Dr.
Fremont, CA 94538
Phone: 510-651-6020
Fax: 510-651-0936
www.bayareaconcretes.com

BOGGS CONTRACTING GROUP, INC.
810 Tipton
Bossier City, LA 71111
Phone: 318-747-3322
Fax: 318-747-6246
www.boggscontracting.com

CONCRETE FX
5737 Kanan Rd., Ste. 283
Agoura Hills, CA 91301
Phone: 818-865-1198
Fax: 818-865-8838
www.concretefx.net

THE CONCRETE IMPRESSIONIST
678 Berriman St.
Brooklyn, NY 11208
Phone: 718-677-1298
Fax: 718-677-0651
www.concreteimpressionist.com

COLORADO HARDSCAPES
8085 E. Harvard Ave.
Denver, CO 80231
Phone: 303-750-8200
Fax: 303-750-8886
www.coloradohardscapes.com

ELITE CONCRETE RESURFACING
1238 Kensington Dr.
High Point, NC 27262
Phone: 336-202-2034
Fax: 336-883-7064
www.eliteconcreteresurfacing.com

EVERLAST CONCRETE
3235 Union Ave.
Steger, IL 60475
Phone: 708-755-0160
Fax: 708-755-0140
www.everlastconcrete.com

HERITAGE BOMANITE
5651 East Fountain Way
Fresno, CA 93727
Phone: 559-291-0506
Fax: 559-291-0515
www.heritagebomanite.com

RICHARD SMITH CUSTOM CONCRETE
6520 Platt Ave., Ste. 257
West Hills, CA 91307
Phone: 818-710-6615
Fax: 818-710-1803
www.richardsmithconcrete.com

SKOOKUM FLOOR CONCEPTS, LTD.
800 Fifth Ave., Ste. 101-411
Seattle, WA 98104
Phone: 206-405-3700
Fax: 206-405-3701
www.concrete-design.com

Bob Harris Decorative Concrete Collection

ALSO AVAILABLE BY THE AUTHOR

www.BobHarrisGuides.com

Bob Harris' Guide to Stained Concrete Interior Floors provides 100-pages of full color, step-by-step instruction for staining concrete floors including specialty methods for applying stains, such as using concrete dyes, sandblasting or creating faux finishes, the tools needed and where to get them, how to maintain concrete floors, and creating intriguing effects with torn paper, cheesecloth, fertilizer, sawdust or kitty litter.

Price $35.00   ISBN: 0-9747737-0-0

Bob Harris' Guide to Stamped Concrete is a 144-page, full color step-by-step guide that shows contractors where to find good designs for stamping concrete, the five methods of imparting color to stamped concrete work, placing the concrete to facilitate stamping, and the stamping process from A-Z.

Price $45.00   ISBN: 0-9747737-1-9